Research on Intelligent Manufacturing

Research on Intelligent Manufacturing (RIM) publishes the latest developments and applications of research in intelligent manufacturing—rapidly, informally and in high quality. It combines theory and practice to analyse related cases in fields including but not limited to:

Intelligent design theory and technologies
Intelligent manufacturing equipment and technologies
Intelligent sensing and control technologies
Intelligent manufacturing systems and services

This book series aims to address hot technological spots and solve challenging problems in the field of intelligent manufacturing. It brings together scientists and engineers working in all related branches from both East and West, under the support of national strategies like Industry 4.0 and Made in China 2025. With its wide coverage in all related branches, such as Industrial Internet of Things (IoT), Cloud Computing, 3D Printing and Virtual Reality Technology, we hope this book series can provide the researchers with a scientific platform to exchange and share the latest findings, ideas, and advances, and to chart the frontiers of intelligent manufacturing.

The series' scope includes monographs, professional books and graduate textbooks, edited volumes, and reference works intended to support education in related areas at the graduate and post-graduate levels.

More information about this series at http://www.springer.com/series/15516

Guilin Yang · I-Ming Chen

Modular Robots: Theory and Practice

華中科技大學出版社
Huazhong University of Science and Technology Press

Springer

Guilin Yang
Ningbo Institute of Materials Technology and Engineering
Chinese Academy of Sciences (CAS)
Ningbo, China

I-Ming Chen
School of Mechanical and Aerospace Engineering
Nanyang Technological University (NTU)
Singapore, Singapore

ISSN 2523-3386 ISSN 2523-3394 (electronic)
Research on Intelligent Manufacturing
ISBN 978-981-16-5006-2 ISBN 978-981-16-5007-9 (eBook)
https://doi.org/10.1007/978-981-16-5007-9

Jointly published with Huazhong University of Science and Technology Press
The print edition is not for sale in China (Mainland). Customers from China (Mainland) please order the print book from: Huazhong University of Science and Technology Press.
ISBN of the China (Mainland) edition: 978-7-5680-7364-6

This Springer imprint is published by the registered company Springer Nature Singapore Pte Ltd.
The registered company address is: 152 Beach Road, #21-01/04 Gateway East, Singapore 189721, Singapore

Preface

A Modular Reconfigurable Robot System (MRRS) consists of an inventory of joint link modules that can be assembled into various robot configurations for a diversity of task requirements. Compared to the conventional robot systems with fixed configurations, such an MRRS has the advantages of reconfigurability, reusability, versatility, and ease of maintenance. Therefore, it has attracted extensive attention from both academic and industrial communities. A variety of MRRSs has been developed for both manufacturing and service applications.

In recent years, the research and development of the MRRSs have been extended from serial robots to parallel robots, especially to those symmetrical parallel robots. Nevertheless, the modular and reconfigurable characteristics of the MRRS bring in a lot of challenges in its kinematics and dynamics modeling, performance analysis, configuration optimization, kinematic calibration, and application developments. To tackle the unfixed-configuration issue of the MRRS, it is critical to investigate configuration-independent modeling and analysis algorithms.

This book is a comprehensive presentation of the work accomplished by the authors in the area of MRRSs, which includes nine chapters. Chapter 1 presents a brief introduction to MRRSs as well as the organization of this book. Chapter 2 introduces the fundamental design issues of robot modules, such as their design requirements, conceptual designs, and geometrical representations. Chapter 3 proposes a kinematic graph-based representation technique, termed Assembly Incidence Matrix (AIM), to mathematically represent modular robot configurations, which lays down a solid foundation for automatic model generations. Chapter 4 addresses the kinematics issues of modular serial robots. A configuration-independent geometric modeling method based on the local presentation of Product-of-Exponentials (POE) formula is employed to automatically generate the forward and inverse kinematic models for general three-structured modular robots. Chapter 5 addresses the kinematic calibration issues of modular serial robots. Based on the local presentation of the POE formula, a configuration-independent kinematic calibration modeling method is proposed for general three-structured modular robots. Chapter 6 addresses the dynamics issues of modular serial robots. A configuration-independent dynamic modeling method based on the Newton-Euler equation is employed to automatically formulate the dynamic models of general three-structured modular robots in both recursive and closed forms. Chapter 7 proposes a task-oriented configuration optimization method for modular serial robots. A Minimized Degrees of Freedom (MDOF) approach is introduced to optimize the robot configurations for a given task. With the virtual module concept and an effective coding scheme, an Evolutionary Algorithm (EA) is employed to conduct both type and dimension optimization for modular serial robots uniformly. Chapter 8 addresses the kinematics issues of modular parallel robots. The local presentation of the POE formula is employed to formulate the displacement, instantaneous kinematics, singularity, and workspace analysis algorithms for a class of three-legged modular parallel robots. Chapter 9 addresses the kinematic calibration issues of modular parallel robots. Based on the local presentation of the POE formula, a two-step kinematic calibration modeling method, i.e., self-calibration followed by base-tool calibration, is proposed for the three-legged modular parallel robots.

By taking tools from graph theory and differential geometry, a unified geometric framework that addresses most of the modeling, analysis, and design issues of the MRRS has been developed by the authors, which is the major feature of this book. A large number of case studies and analysis examples are provided to facilitate the audience to make use of this book for both academic research and practical implementation. It is worth to mentioning that the modeling and analysis methods developed for the MRRS are generic so that they can be applied to conventional serial and parallel robots. Hence, although this book is primarily intended for researchers and engineers working on modular reconfigurable robots, it could also be of interest to readers in a broad area of robotics.

This book would not have been possible without the help of many people. In particular, we would like to thank Prof. Song Huat Yeo from Nanyang Technological University who contributed very useful suggestions to this book. We would also like to thank our graduate students Mr. Wenji Jia, Mr. Renfeng Zhu, Mr. Yi Wang, Ms. Wenjun Shen, Mr. Jie Zhao, Mr. Qinghao Du, Mr. Tuopu Zhang, Mr. Lefeng Gu, Mr. Kaisheng Yang, Mr. Hao Zhang Mr. Guolong Zhang, and Mr. Mengshen Yang for their great assistance in preparing the manuscripts. The research work presented in this book is partially supported by Nanyang Technological University and Singapore Institute of Manufacturing Technology, Singapore, which are dutifully acknowledged. The authors also gratefully acknowledge the continuous financial support of the National Natural Science Foundation of China (NSFC), especially, under the grants U1509202 and 92048201.

Ningbo, China Guilin Yang
Singapore, Singapore I-Ming Chen
March 2021

Contents

List of Figures

List of Tables

Chapter 1
Introduction

1.1 Motivation

Modularity usually refers to the design concept of making multiple independent functional units first and then integrating them into a complete system. Such a design concept has been seen in many engineering disciplines: mechanical engineering, electrical and electronics engineering, and computer and software engineering. Modularity will bring in the capabilities of re-usability, rapid change-over, and ease-of-maintenance. A system designed in such a way would be reconfigurable, scalable, and flexible. From the system level aspect, the overall performance may not be optimized although the individual unit is optimal. More integration effort, for instance, the compatibility between different units, has to be made on modularly designed systems.

Robotic automation systems are extensively employed for manufacturing applications. Nowadays, they also make their ways into other areas, such as medical and health care, hazardous environment exploration, security, military, law enforcement, and entertainment. In the design and selection of an effective robotic automation system for a set of tasks or services, however, system flexibility and productive capacity are always the main concern. Figure 1.1 shows the application scenarios of different robotic and automation systems based on the variety and volume (or quantity) of the tasks (services). For high-volume/low-variety tasks (services), fully customized equipment is preferred because the capital investment can be justified by the reliability and productivity provided by the system. For low-volume/high-variety tasks (services), fully general-purpose machines can be employed because the high profit margin can cover the capital investment and maintenance cost. However, there is a large area of tasks (services) that belongs to the mid-volume/mid-variety range, which usually is a compromise between flexibility and productivity. In this case, employing modularly designed robotic automation systems becomes an attractive solution. This is because (1) the flexibility of the system can be enabled through

G. Yang and I.-M. Chen, *Modular Robots: Theory and Practice*,
Research on Intelligent Manufacturing,
https://doi.org/10.1007/978-981-16-5007-9_1

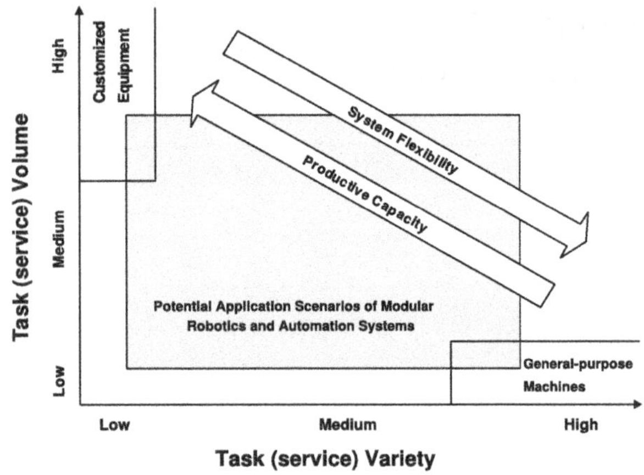

Fig. 1.1 Application scenarios of robotic automation systems

the reconfiguration of the modular units, and at the same time (2) the high level of productivity can be achieved through the optimally designed modular workcells that provide high operational efficiency.

This book mainly discusses issues related to Modular Reconfigurable Robot Systems (MRRSs). An MRRS is a collection of individual link and joint modules that can be assembled into various robot configurations for a diversity of task requirements. Such a modular robotic system has a variety of application scenarios: an individual joint module can be used as a self-contained motion control element; fully assembled systems can be used as robots with a variety of configurations and Degrees of Freedom (DOFs). Here we focus on the fully assembled systems—the modular robots. Many research issues, such as module design, kinematic and dynamic modeling, kinematic calibration, and configuration optimization, arise during the integration of the modules into effective robot systems.

1.2 Past Research and Development Efforts

There has been an increasing interest in designing modularized robot systems in past few decades. The research efforts include system hardware design [1–12], kinematics [4, 13–18], dynamics [15, 19–21], control [8, 10, 22], calibration [15, 23–26], and task-based design of modular robot configurations [7, 14, 22, 27–34]. This section briefly reviews the fundamental research efforts on modular robotics. Literature surveys on the individual topics will be presented in the corresponding chapters.

The most fundamental issue in modular robotics is the system hardware design including both mechanical and electronic designs. Up to date, several prototype systems have been demonstrated in various research institutions, such as the Recon-

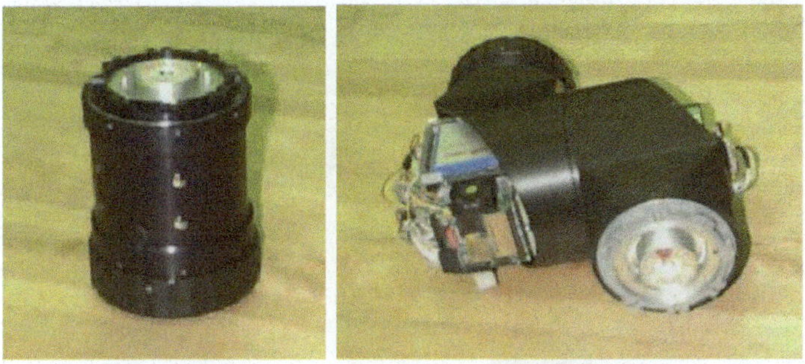

(a) link module (b) pivot joint module

Fig. 1.2 An RMMS link and pivot joint module

figurable Modular Manipulator System (RMMS) at Carnegie Mellon University [9, 10], the Dynamically Reconfigurable Robotic System (DRRS) at the Science University of Tokyo [6], the TOshiba Modular Manipulator System (TOMMS) at Toshiba Corp. [8], and other modular robot systems at University of Toronto [5], University of Stuttgart [12], and University of Texas [1, 11].

The RMMS at Carnegie Mellon University [9, 10, 22] consists of two types of 1-DOF revolute joints: "rotate" and "pivot" joints, and links of different sizes (Fig. 1.2). The revolute joints are actuated by DC motors in conjunction with harmonic-drive mechanisms. The links have cylindrical shapes. The mechanical coupling uses the V-band clamp, which is an integral part of link and joint modules. In the second generation of the RMMS [9, 22], a quick-coupling mechanism is designed for easy module connection. An RMMS robot configuration is shown in Fig. 1.3. The modules are communicated with a VME-based host controller over a local area network called the ARMbus. The control software for the RMMS has been developed using the Chimera real-time operating system. Because closed-form inverse kinematics solutions do not exist for all possible RMMS configurations, a damped least-squares kinematic control algorithm is employed. A number of research issues are addressed extensively for the RMMS: the automatic generation of kinematics [17], the global fault tolerant planning [22, 35], and the task-based design problem [22, 31–33, 36].

The Modular Robotic System (MRS) at University of Toronto [5] includes two types of 1-DOF main joint modules (rotary and prismatic), three types of 1-DOF end-effector joint modules (yaw, pitch, and roll type), and link modules with hollow cylindrical geometry (Fig. 1.4). All the joint modules are designed to be self-contained independent units. A revolute joint module driven by a DC motor with a harmonic-drive mechanism can generate rotary motion; a prismatic joint module driven by a DC motor in conjunction with a ball-screw mechanism can generate linear motion. A 4-DOF SCARA type modular robot configuration constructed with these modules is shown in Fig. 1.5. A unique feature of this system is the development of a 45-degree (relative to the modules' main axes) connection structure. This structure allows both types of straight and perpendicular connections by simply rotating the next module

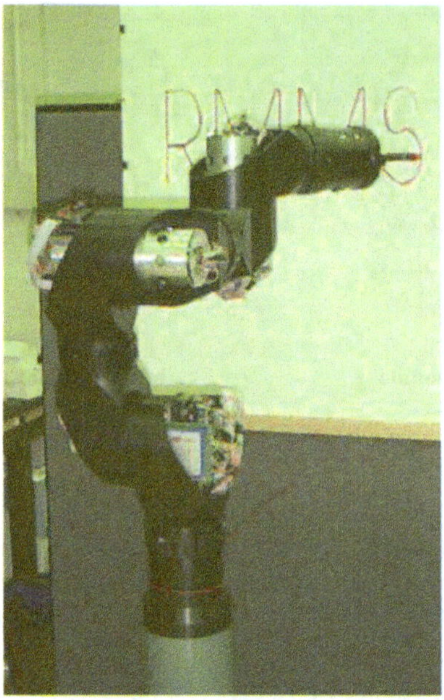

Fig. 1.3 A robot configuration constructed by RMMS modules

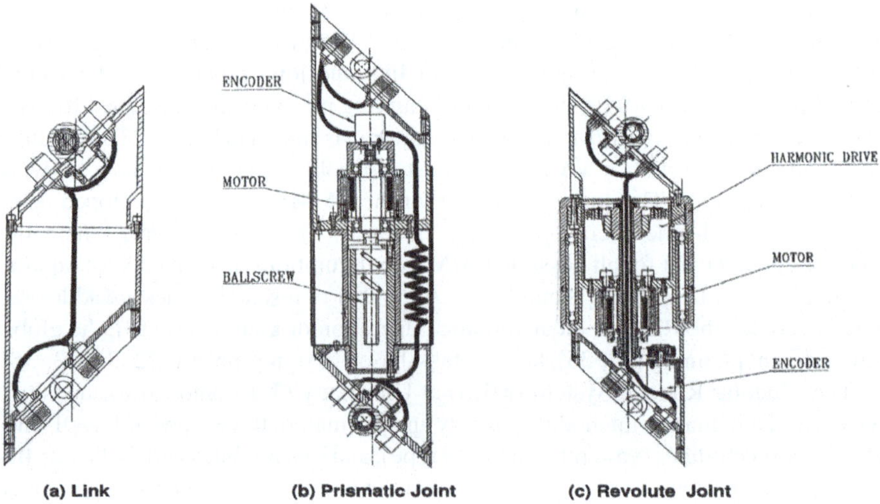

Fig. 1.4 Robot modules in the MRS

by 180° about its main axis. Based on the Denavit–Hartenberg parameter notation, a configuration independent kinematic modeling technique is also proposed for this system [13].

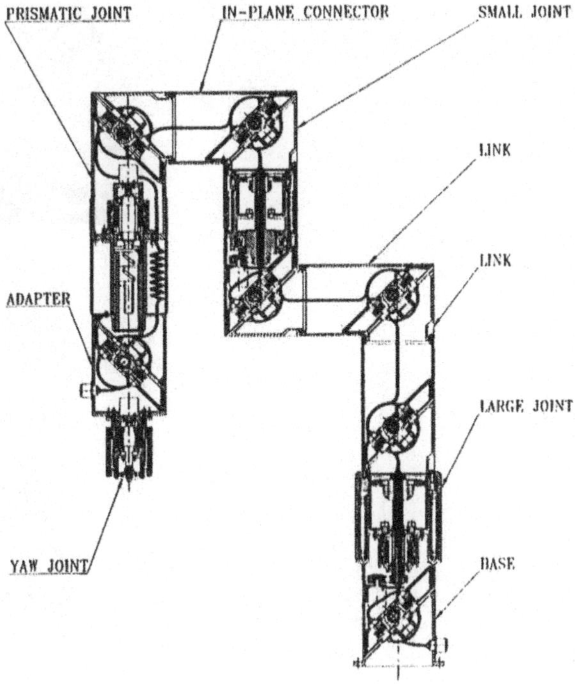

Fig. 1.5 A SCARA type modular robot configuration constructed by MRS modules

The modular robot at University of Stuttgart [12] consists of link modules of different sizes and a variety of rotational joint modules. The joint modules driven by AC motors in conjunction with differential gears can create complicated 2-DOF or 3-DOF motions. A simple male-female fastening device is designed to connect modules in any order quickly and exactly. All the necessary specifications of link and joint modules are stored in a database file. A program that can select modules to construct special-purpose modular manipulators is proposed.

The TOMMS at Toshiba Corp. [8] contains one type of 1-DOF revolute joint module actuated by a DC motor with harmonic gears, and one type of link module. The joint module has three input ports and two output ports. The three input ports are used for the alignment of the rotational axis with three axes of the joint coordinate system respectively, while the two output ports are used to permit the module to either rotate or bend. The link module consists of two cylinders that can slide relative to each other so that the length of the link module is adjustable. Based on numerical inverse kinematics, a position control algorithm is designed for the TOMMS.

There are also many research efforts dedicated to the modular joints designs and applications, such as the collaborative robots. Some research groups have developed modular joint prototypes, the modular joint integrates electronics and mechanical components, which includes high efficiency and high power density Permanent Magnet Synchronous Motor (PMSM), harmonic drive, dual encoders (motor-side

Fig. 1.6 The modular joints developed in NIMTE

Fig. 1.7 M-TRAN

encoder and link-side encoder), torque sensor, and other components, as shown in Fig. 1.6.

In the early 2000s, the modular self-reconfigurable robots start to draw the attention of many researchers [37–41]. Modular self-reconfigurable robots involve various modules that can combine themselves autonomously and make transformations into many different topological configurations that are capable of performing various tasks. The ability to self-reconfigure allows these robots of metamorphosis, which in turn makes them capable of performing different sorts of kinematics. For instance, a robot may transform into a multi-leg configuration to walk through uneven terrain, and then re-transform into a snake-like robot to move through a cave. This sort of adaptability enables self-reconfigurable robots to accomplish tasks in unstructured environments; such as space exploration, deep sea applications, rescue missions, or reconnaissance. Some representative modular self-reconfigurable robots are: M-TRAN (Fig. 1.7) [42], ATRON (Fig. 1.8) [43], SUPERBOT (Fig. 1.9) [44], SMORSE (Fig. 1.10) [45].

An extensive study on modular robot mechanical structures is performed in the University of Texas at Austin [1, 2, 11]. An inventory of joint modules including 1-DOF "elbows", 2-DOF "knuckle", and 3-DOF "shoulders" and "wrists", and a variety of link modules with different lengths, cross-sections, and materials are designed and tested. To evaluate the performance of an individual module, two performance matrices, namely, joint module performance matrix and link module performance matrix are defined. This research actually advocates a "component" design

Fig. 1.8 ATRON

Fig. 1.9 SUPERBOT

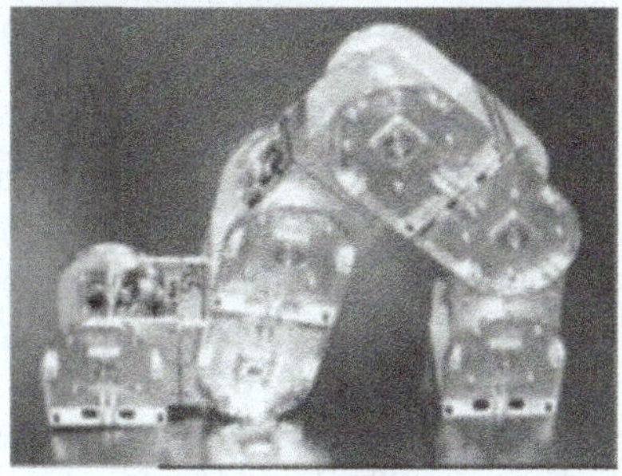

Fig. 1.10 SMORSE

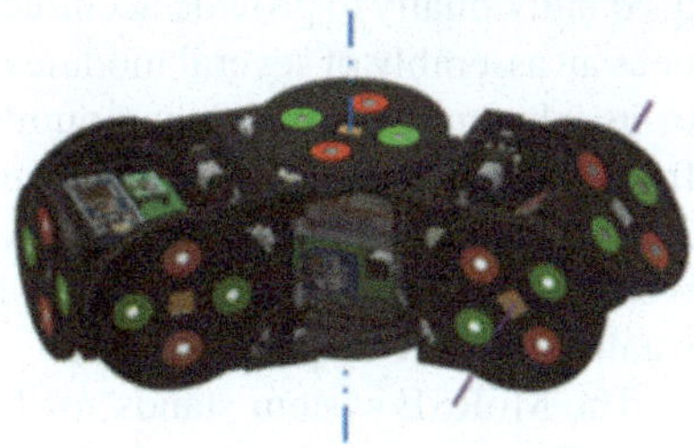

approach, rather than a general "complete structure" design approach. Several optimization algorithms are also proposed for the simple (one and two-dimensional) robot configuration design [1].

The DRRS at the Science University of Tokyo [6, 46], however, has a very different system design. The idea of this study is to apply the cell system of living creatures to a robot system with some intelligence analogous to the biological gene. There are four types of cells (modules) in the DRRS: joint cells, mobile cells, branching cells, and work cells. All cells are identical in dimensions. The connection and detachment between cells are carried out by the mobile or joint cells through a hook type coupling mechanism. A cone-shaped mechanism guides the cells during the coupling. The cells in the DRRS possess certain intelligence and communication capability so that they can reconfigure themselves autonomously.

In addition to the research types of MRSs, there are a few commercial MRSs: RMD-1 by Engineering Services Inc. of Canada and MoRSE by AMTEC GmbH of Germany.

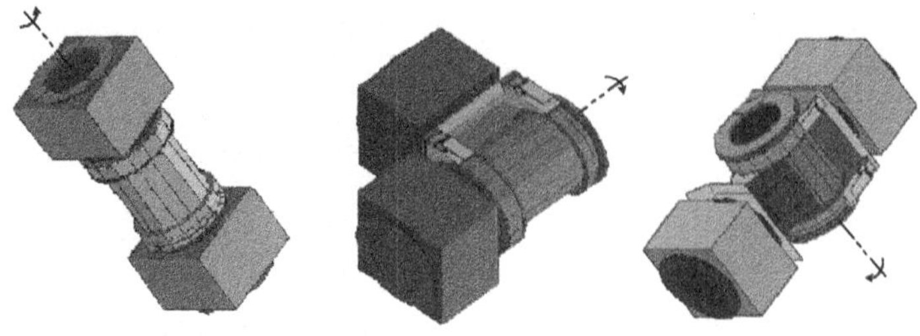

Fig. 1.11 Three different types of rotations

The RMD-1 is a reconfigurable and modular (rotary) mechatronic drive system (joint) (Fig. 1.11). Each RMD-1 comprises a built-in actuator, a gear mechanism, a position encoder, a brake, all of the necessary electronics, and the on-board processing unit. RMD-1 is available in various power capacities starting from 50 W to about 400 W. The size of RMD-1 varies with its power capacity. Each RMD-1 module can be reconfigured to provide three different types of rotations: roll, pitch, and yaw. Serving as a "building block" of motion control automation, RMD-1 can either be used individually to provide accurate motion control of a payload about a single axis, or as an assembly of several modules to provide control about multiple axes. Because of its inherent modularity, any number of modules can be added or removed easily from an existing configuration. Figure 1.12 shows two modules interconnected in a 2-DOF configuration. And a modular and reconfigurable robot system built by ESI is shown in Fig. 1.13. Note that both Figs. 1.11 and 1.12 are adapted from the product catalog of ESI Company, Canada.

The MoRSE system stands for Modular Robot System for Engineering. It consists of a variety of drive modules of different sizes: 1-DOF revolute drive modules, 1-DOF linear drive modules, 2-DOF wrist modules, and gripper modules (Fig. 1.14). Each MoRSE module is a totally self-contained unit. The brushless DC servo motor, the motor controller, the harmonic-drive mechanism, the break, the opto-isolated digital and analog I/O, and the Fieldbus (CAN bus) interface are all encapsulated in a compact design. Due to the standardized cubic design, the modules of MoRSE system can be connected via simple coupling elements (connectors). An existing robot configuration can be easily altered when required. A PUMA type robot configuration constructed by using MoRSE drive modules is shown in Fig. 1.15. Note that both Figs. 1.14 and 1.15 are adapted from the product catalog of AMTEC GmbH, Germany.

Another Germany company SCHUNK, has launched a commercial MRRS named Powerball LWA 4 (Fig. 1.16). The SCHUNK Powerball LWA 4 is a compact lightweight arm that utilizes all the advantages of the SCHUNK mechatronic modular system. Three ERB modules can achieve up to six degrees of freedom. The ERB modules combine the movements of two servo axes per module in a very compact

Fig. 1.12 A 2-DOF RMD-1 assembly configuration

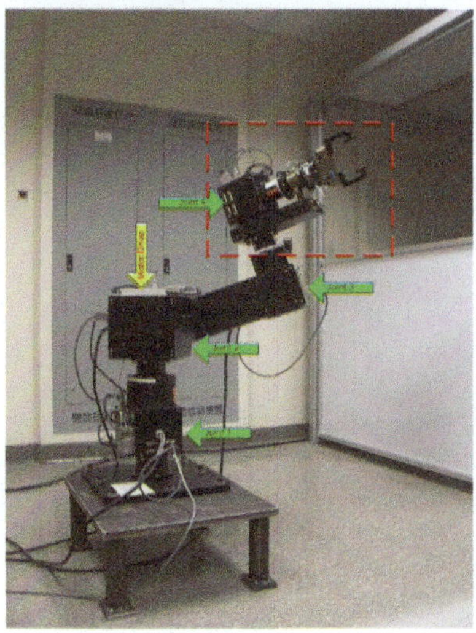

Fig. 1.13 Modular and reconfigurable robot system built by ESI

design. The combination of different sizes results in a customized, flexible manipulator for modern handling concepts with a payload of up to 3 kg.

Based on the existing MRSs, the conceptual design of an MRRS has been proposed in our early work [27]. It contains most of the features of the existing systems. An individual joint module can be used as an independent motion control element. The fully assembled systems can be used as standard robots with various configurations and DOFs. The modules making up the MRRS can be divided into two categories,

Fig. 1.14 MoRSE drive modules

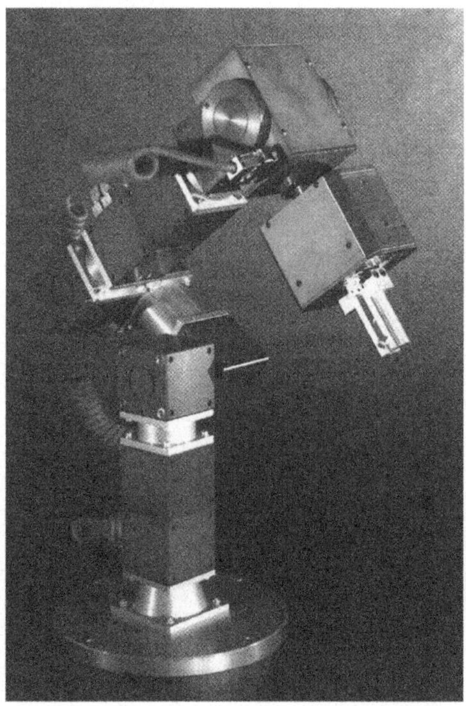

Fig. 1.15 A PUMA type robot configuration constructed by MoRSE modules

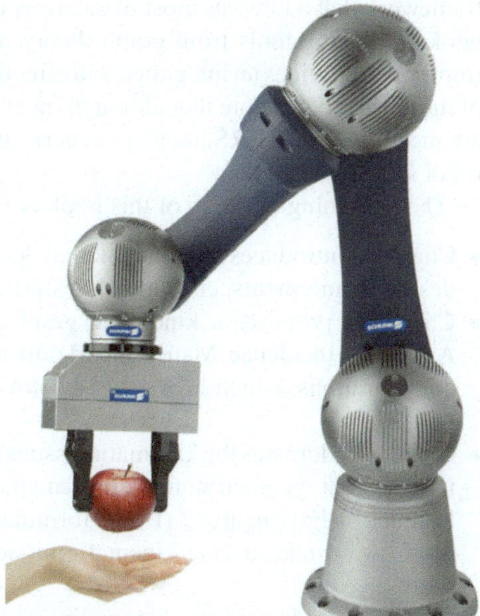

Fig. 1.16 SCHUNK Powerball LWA 4

i.e., joint modules and link modules. The joint modules are self-contained control units, including 1-DOF revolute and prismatic joint modules. The connectors for mechanical and electrical coupling of modules have also been designed. Based on the building block principle, each module is designed as a cube with multiple connecting sockets so that many possible configurations can be constructed, e.g., serial and tree-structured types. A detailed introduction to the MRRS is addressed in Chap. 2.

Note that past research efforts on modular robot systems are mainly focused on the hardware design. Various types of robot modules and module connectors have been designed and developed. The software aspect of modular robots, however, is yet to be studied intensively. Algorithms, such as kinematic and dynamic modeling algorithms, employed in most of the existing modular robot systems are transplanted from the conventional robot systems and are configuration-dependent. Module and configuration selection and optimization algorithms are in their primitive stages. Hence, in order to develop an effective MRRS, the algorithm and software aspect of modular robots should be emphasized.

1.3 Overview of This Book

This book addresses several fundamental research topics relating to the MRRS, i.e., conceptual module design, module assembly representation, kinematics, dynamics, calibration, and optimization of modular robot configurations. A unified geometric framework that addresses most of the aspects of the MRRS has been proposed in this book. By taking tools from graph theory and differential geometry, the geometric framework provides an integrated and effective tool for modeling, analysis and design of modular robots. Note that although most of the algorithms developed in this book are meant for the MRRS, they are generic and can be applied to all existing modular robot systems as well.

The remaining chapters of this book are organized as follows:

- Chapter 2 introduces the fundamental design issues of robot modules, such as their design requirements, conceptual designs, and geometrical representations.
- Chapter 3 proposes a kinematic graph based representation technique, termed Assembly Incidence Matrix (AIM), to mathematically represent modular robot configurations, which lays down a solid foundation for automatic model generations.
- Chapter 4 addresses the kinematics issues of modular serial robots. A configuration-independent geometric modeling method based on the local presentation of Product-of-Exponentials (POE) formula is employed to automatically generate the forward and inverse kinematic models for general three-structured modular robots.
- Chapter 5 addresses the kinematic calibration issues of modular serial robots. Based on the local presentation of the POE formula, a configuration-independent kinematic calibration modeling method is proposed for general tree-structured modular robots.
- Chapter 6 addresses the dynamics issues of modular serial robots. A configuration independent dynamic modeling method based on the Newton–Euler equation is employed to automatically formulate the dynamic models of general tree-structured modular robots in both recursive and closed forms.
- Chapter 7 proposes a task-oriented configuration optimization method for modular serial robots. A Minimized Degrees of Freedom (MDOF) approach is introduced to optimize the robot configurations for a given task. With the virtual module concept and an effective coding scheme, an Evolutionary Algorithm (EA) is employed to conduct both type and dimension optimization for modular serial robots uniformly.
- Chapter 8 addresses the kinematics issues of modular parallel robots. The local presentation of the POE formula is employed to formulate the displacement, instantaneous kinematics, singularity, and workspace analysis algorithms for a class of three-legged modular parallel robots.
- Chapter 9 addresses the kinematic calibration issues of modular parallel robots. Based on the local presentation of the POE formula, a two-step kinematic calibration modeling method, i.e., self-calibration followed by base-tool calibration, is proposed for the three-leg modular parallel robots.

Chapter 2
Module Designs

Abstract This chapter introduces the conceptual designs of both joint and link modules. Although the actual joint and link modules may have different geometrical shapes, these modules are often connected with each other either in co-linear or orthogonal ways. Therefore, to generically describe the connectivity of a Modular Reconfigurable Robot System (MRRS), the joint and link modules are conceptually designed as cubes or rectangular cuboids so that multiple faces of a joint or link module can be used as the connecting interfaces with other modules.

2.1 Module Design Requirements

In order to develop a Modular Reconfigurable Robot System (MRRS) with versatile reconfigurability, the module design in an MRRS should satisfy the following design requirements:

- Each joint module should be a self-contained and independent actuator with its own motion control capabilities;
- Each joint or link module should be designed with multi-connectivity;
- Each joint or link module should be designed with minimum weight and inertia;
- The connections between modules should be precise and rigid.

The first requirement focuses on the reconfigurability of the system, which implies that the joint module should have built-in servo-motor, gearbox, amplifier, controller, encoder, brake, and communication interface. The second requirement is the essential embodiment to increase the reconfigurability of the system. When each type of the module is designed with multiple connecting sockets, many possible robot configurations including serial, three-structured, and parallel types can be constructed. The third requirement is important for a modular robot to increase its loading capac-

G. Yang and I.-M. Chen, *Modular Robots: Theory and Practice*,
Research on Intelligent Manufacturing,
https://doi.org/10.1007/978-981-16-5007-9_2

ity. The last requirement is essential for a modular robot to perform tasks requiring high accuracy and rigidity.

2.2 Joint Modules

There are five types of lower pair joints commonly used in robot systems: revolute (1-DOF), prismatic (1-DOF), cylindrical (2-DOF), planar (2-DOF), and spherical (3-DOF) [14]. Since a multi-DOF joint can be obtained by combining several 1-DOF joints, it is unnecessary to have complex multi-DOF joint designs in the MRRS. Hence, only two types of 1-DOF joint modules, i.e., revolute joint modules and prismatic joint modules are considered.

2.2.1 Revolute Joint Modules

A revolute joint module is designed to generate a rotary motion. To generalize the revolute joint module, it is conceptually designed as a cube, as shown in Fig. 2.1. The actuator is packed within the joint housing. Among its six connecting sockets, one is the revolute socket driven by the built-in actuator. For example, in order to get a high torque/weight ratio within a compact design, the commercial FH-2000 series hollow-shaft servo actuators (from Harmonic Drives GmbH, Germany) can be selected for the design, as shown in Fig. 2.2. Such an actuator combines a maintenance-free AC servo motor with an encoder, a backlash-free harmonic drive, and a brake in a compact design. The unique feature of the actuators is the hollow shaft which can allow the electrical wiring to pass through the shaft.

For the conceptual revolute joint module (Fig. 2.3), note that the term 'CM' refers to the center of mass. The location of CM is expressed in the module frame which is located at the center of the cube as shown in Fig. 2.3. The initial matrix is given with respect to the frame at CM, termed as CM frame. The CM frame is parallel with the module frame.

2.2.2 Prismatic Joint Modules

A prismatic joint module is designed to generate a translational motion. Such a module is conceptually designed to have the similar multiple connecting sockets on its housing as the revolute joint modules, as shown in Fig. 2.4. One of them is the translational socket driven by an actuator in conjunction with a ball-screw transmission mechanism. The FH series AC servo motors (without harmonic drives) can also be used as the actuators of the prismatic joint modules for the same reason mentioned above.

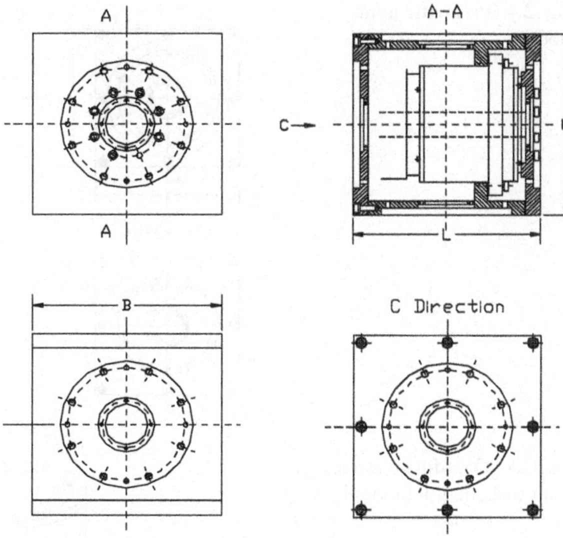

Fig. 2.1 Revolute joint module

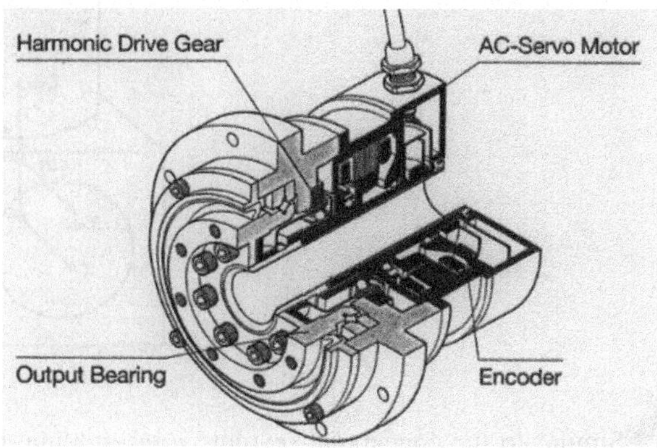

Fig. 2.2 FH-2000 series motor

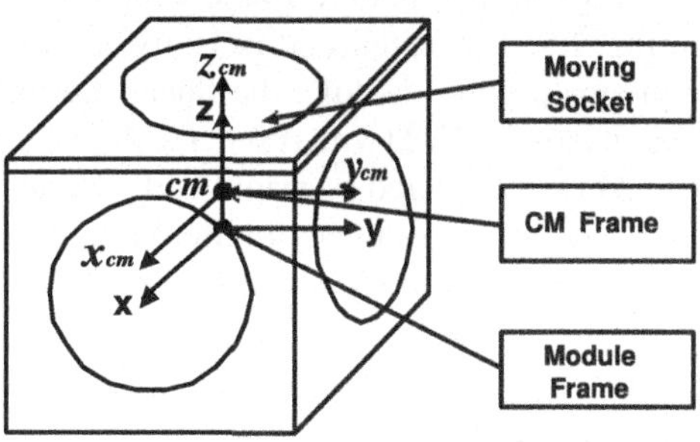

Fig. 2.3 Coordinate frames in a revolute joint module

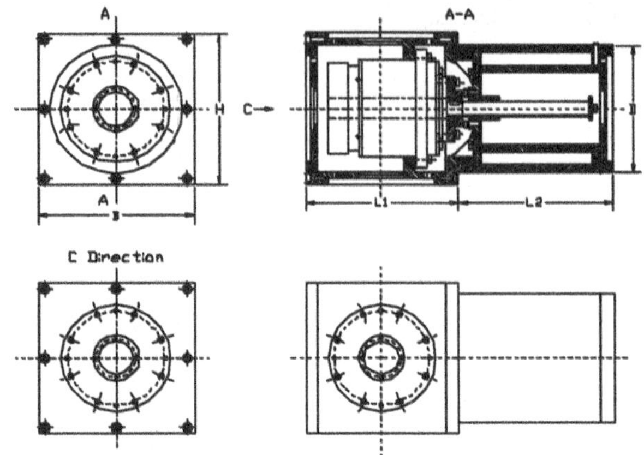

Fig. 2.4 Prismatic joint module

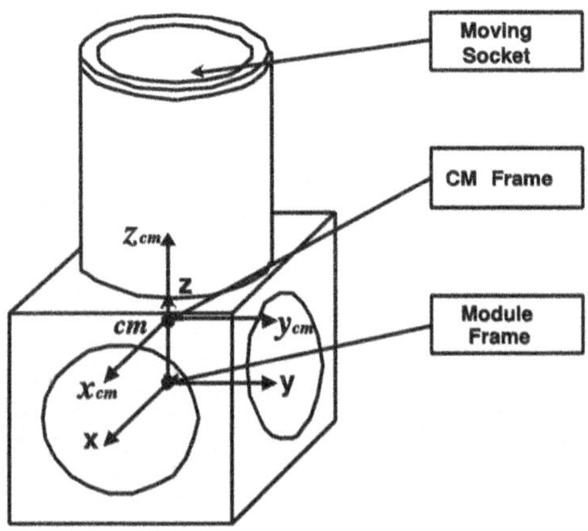

Fig. 2.5 Coordinate frames in a prismatic joint module

Similar to the conceptual revolute joint module design, the location of CM is expressed in the module frame which is located at the center of the cube as shown in Fig. 2.5. The initial matrix is given with respect to the CM frame. Note that for a prismatic joint module, the linear motion of the cylindrical tube will change the position of the CM (with respect to the module frame). In this case, the whole module cannot be considered as a rigid body. Hence, we divide a prismatic joint module into two parts: the movable cylindrical tube and the remaining part, which is still called a prismatic joint module without confusion.

2.3 Link Modules

Link modules are used to connect or support joint modules by means of connectors. The geometry of the link module is similar to the revolute joint module but without the

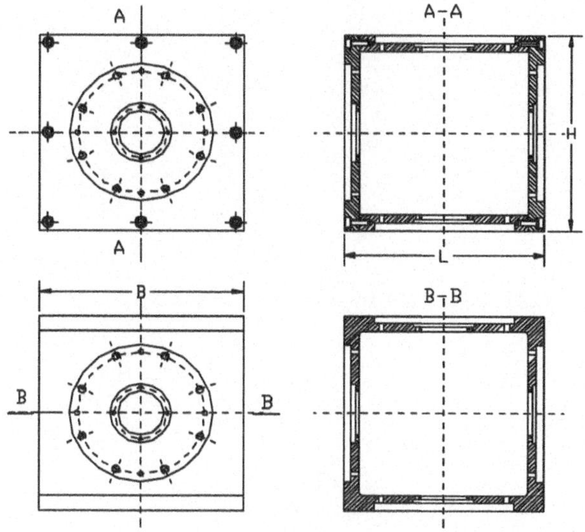

Fig. 2.6 Cubic link module

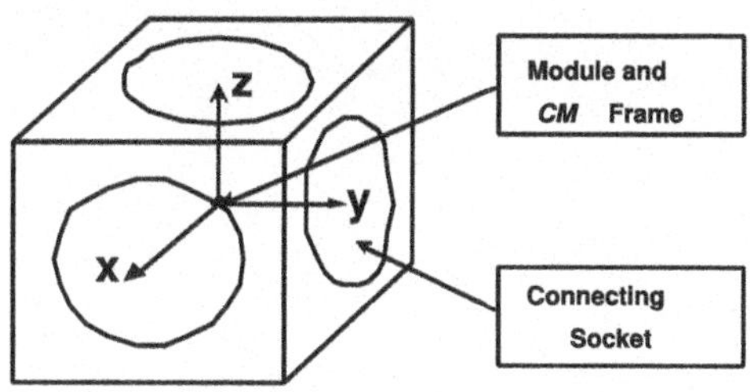

Fig. 2.7 Coordinate frames in cubic link module

built-in actuator and the moving socket. In the conceptual design of the link modules, the six connecting sockets in a link module are all fixed as shown in Fig. 2.6. The location of CM is expressed in the module frame which is located at the center of the cube, as shown in Fig. 2.7. The initial matrix is given with respect to the CM frame. Because the link module is a symmetric cube, the model frame and the CM frame coincide with each other.

Chapter 3
Modular Robot Representation

Abstract This chapter reviews the basic concepts, definitions, and algorithms in graph theory, which is important to the mathematical representation of modular robot assembly configurations. Since a modular robot is a collection of link and joint modules, a re-classification scheme for links and joints is proposed to translate a modular robot into a kinematic graph. Finally, the modified Assembly Incidence Matrix (AIM) is introduced to fully describe the assembly configuration of a modular robot in a compact matrix form, which makes the automatic generation of the kinematic, dynamic, and calibration models more convenient.

3.1 Graphs

In this section, several basic concepts and definitions in graph theory are reviewed. All the graphs discussed here are simple graphs. For more details, please refer to [48, 49].

3.1.1 Basic Graph Definitions

Definition 3.1 (*Graph*) A graph $G = (V, E)$ consists of a vertex set, $V(G)$, and an edge set, $E(G)$, such that every edge in $E(G)$ is associated with a pair of vertices in $V(G)$ (Fig. 3.1a).

Definition 3.2 (*Directed Graph*) A directed graph (or digraph) is a graph in which every edge is associated with an ordered pair of vertices (Fig. 3.1b).

Definition 3.3 (*Connected Graph*) A graph is connected if for every pair of distinct vertices $\{v_i, v_j\}$ there is a path from v_i to v_j (Fig. 3.2a).

G. Yang and I.-M. Chen, *Modular Robots: Theory and Practice*,
Research on Intelligent Manufacturing,
https://doi.org/10.1007/978-981-16-5007-9_3

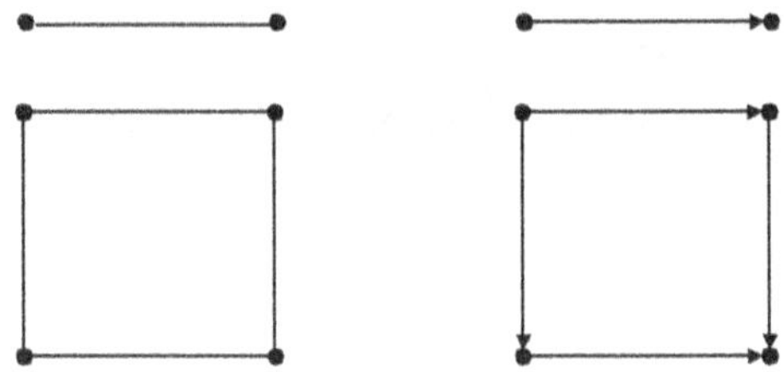

Fig. 3.1 A graph and a directed graph

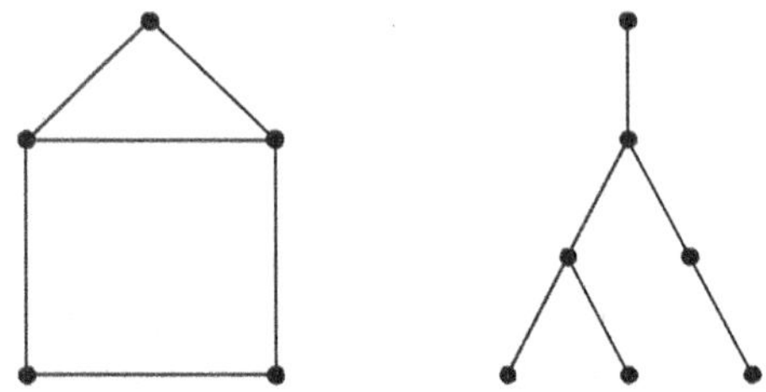

Fig. 3.2 A connected graph and a tree

Definition 3.4 (*Tree*) A tree is a (minimal) connected graph without cycle (Fig. 3.2b).

Note that, for a tree, if the root is given, the tree will become a directed or rooted tree. Moreover, a tree with n vertices has $n - 1$ edges.

Definition 3.5 (*Labeled Graph*) A labeled graph is a graph in which the vertices are labeled by $\mathrm{v}_1, \mathrm{v}_2, \ldots, \mathrm{v}_\mathrm{m}$, and the edges are labeled by $\mathrm{e}_1, \mathrm{e}_2, \ldots, \mathrm{e}_\mathrm{n}$, such that $V = \{\mathrm{v}_1, \mathrm{v}_2, \ldots, \mathrm{v}_\mathrm{m}\}$ and $E = \{\mathrm{e}_1, \mathrm{e}_2, \ldots, \mathrm{e}_\mathrm{n}\}$ (Fig. 3.3a).

Usually, we do not differentiate among the types of vertices (edges). If the vertices (edges) are assumed to be different, a more complex graph, termed specialized graph, should be defined.

Definition 3.6 (*Specialized Graph* [14]) A specialized graph $\tilde{G}$ is a labeled graph G with an assignment (f_v, f_e) on its vertices and edges. It can be written as $\tilde{G} = (G, \tilde{V}, \tilde{E}, f_v, f_e)$, where $\tilde{V}(\tilde{E})$ represents the set of vertex (edge) types on G, and $f_v : V \mapsto \tilde{V}$ and $f_e : E \mapsto \tilde{E}$ are injective mappings which are called vertex and edge assignments, respectively (Fig. 3.3b).

Example 3.1 A labeled graph G is shown in Fig. 3.3a, where $V = (\mathrm{v}_1, \mathrm{v}_2, \ldots, \mathrm{v}_6)$ and $E = (\mathrm{e}_1, \mathrm{e}_2, \ldots, \mathrm{e}_5)$. Its specialized graph $\tilde{G}$ is shown in Fig. 3.3b, where $\tilde{V} = \{\mathrm{A, B, C}\}$ and $\tilde{E} = \{\mathrm{x, y, z}\}$.

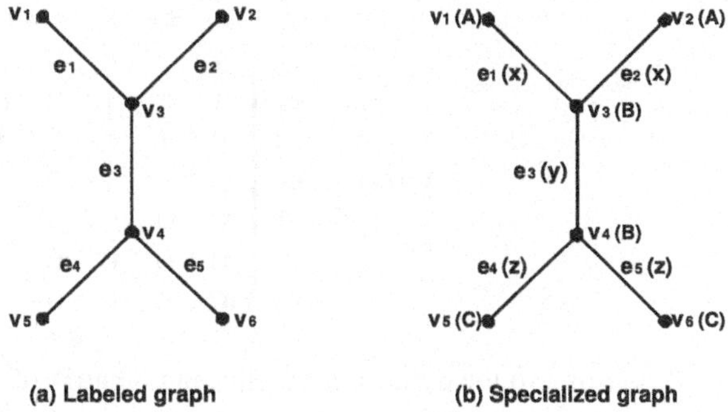

Fig. 3.3 A labeled graph and a specialized graph

3.1.2 *Matrix Representation of Graphs*

A graph can also be expressed in matrix forms that are very convenient for computer processing. The following matrix representations for graphs are extensively used throughout this book.

Definition 3.7 (*Adjacency Matrix*) Let $G = (V, E)$ be a labeled graph, where $V = \{v_1, v_2, \ldots, v_m\}$ and $E = \{e_1, e_2, \ldots, e_n\}$. The adjacency matrix $\mathcal{B}(G)$ is the $m \times m$ matrix in which the entry in row i and column j is 1, if there is an edge joining vertices v_i and v_j; it is 0, otherwise.

Definition 3.8 (*Incidence Matrix*) Let $G = (V, E)$ be a labeled graph, where $V = \{v_1, v_2, \ldots, v_m\}$ and $E = \{e_1, e_2, \ldots, e_n\}$. The incidence matrix $\mathcal{M}(G)$ is the $m \times n$ matrix in which the entry in row i and column j is 1, if edge e_j is incident on vertex v_i; it is 0, otherwise.

Example 3.2 The adjacency and incidence matrix of the underlying graph shown in Fig. 3.4a can be written as:

$$
\mathcal{B}(G) = \begin{array}{c} \\ v_0 \\ v_1 \\ v_2 \\ v_3 \\ v_4 \\ v_5 \\ v_6 \end{array}
\begin{array}{c} \begin{array}{ccccccc} v_0 & v_1 & v_2 & v_3 & v_4 & v_5 & v_6 \end{array} \\
\begin{pmatrix}
0 & 1 & 0 & 0 & 0 & 0 & 0 \\
1 & 0 & 1 & 1 & 0 & 0 & 0 \\
0 & 1 & 0 & 0 & 1 & 1 & 0 \\
0 & 1 & 0 & 0 & 0 & 0 & 1 \\
0 & 0 & 1 & 0 & 0 & 0 & 0 \\
0 & 0 & 1 & 0 & 0 & 0 & 0 \\
0 & 0 & 0 & 1 & 0 & 0 & 0
\end{pmatrix} \end{array},
$$

$$\mathcal{M}(G) = \begin{array}{c} \\ \mathbf{v}_0 \\ \mathbf{v}_1 \\ \mathbf{v}_2 \\ \mathbf{v}_3 \\ \mathbf{v}_4 \\ \mathbf{v}_5 \\ \mathbf{v}_6 \end{array} \begin{array}{c} \begin{array}{cccccc} \mathbf{e}_1 & \mathbf{e}_2 & \mathbf{e}_3 & \mathbf{e}_4 & \mathbf{e}_5 & \mathbf{e}_6 \end{array} \\ \begin{pmatrix} 1 & 0 & 0 & 0 & 0 & 0 \\ 1 & 1 & 1 & 0 & 0 & 0 \\ 0 & 1 & 0 & 1 & 1 & 0 \\ 0 & 0 & 1 & 0 & 0 & 1 \\ 0 & 0 & 0 & 1 & 0 & 0 \\ 0 & 0 & 0 & 0 & 1 & 0 \\ 0 & 0 & 0 & 0 & 0 & 1 \end{pmatrix} \end{array}.$$

According to the definition of adjacency matrix of an underlying graph, $\mathcal{B}(G)$ is a square and symmetric matrix.

Definition 3.9 (*Adjacency Matrix—for a directed graph*) Let $\mathbf{G} = (V, E)$ be a labeled digraph, where $V = \{\mathrm{v}_1, \mathrm{v}_2, \ldots, \mathrm{v}_\mathrm{m}\}$ and $E = \{\mathrm{e}_1, \mathrm{e}_2, \ldots, \mathrm{e}_\mathrm{n}\}$. The adjacency matrix $\mathcal{B}(\mathbf{G})$ is the $m \times m$ matrix in which the entry in row i and column j is 1, if there is an edge from vertices v_i to v_j; it is 0, otherwise.

Definition 3.10 (*Incidence Matrix—for a directed graph*) Let $\mathbf{G} = (V, E)$ be a labeled digraph, where $V = \{\mathrm{v}_1, \mathrm{v}_2, \ldots, \mathrm{v}_\mathrm{m}\}$ and $E = \{\mathrm{e}_1, \mathrm{e}_2, \ldots, \mathrm{e}_\mathrm{n}\}$. The incidence matrix $\mathcal{M}(\mathbf{G})$ is the $m \times n$ matrix in which the entry in row i and column j is 1 if v_i is the initial vertex of e_j; it is -1, if v_i is the terminal vertex of e_j; it is 0, otherwise.

Example 3.3 The adjacency and incidence matrix of the labeled digraph shown in Fig. 3.4b can be written as:

$$\mathcal{B}(\mathbf{G}) = \begin{array}{c} \\ \mathbf{v}_0 \\ \mathbf{v}_1 \\ \mathbf{v}_2 \\ \mathbf{v}_3 \\ \mathbf{v}_4 \\ \mathbf{v}_5 \\ \mathbf{v}_6 \end{array} \begin{array}{c} \begin{array}{ccccccc} \mathbf{v}_0 & \mathbf{v}_1 & \mathbf{v}_2 & \mathbf{v}_3 & \mathbf{v}_4 & \mathbf{v}_5 & \mathbf{v}_6 \end{array} \\ \begin{pmatrix} 0 & 1 & 0 & 0 & 0 & 0 & 0 \\ 0 & 0 & 1 & 1 & 0 & 0 & 0 \\ 0 & 0 & 0 & 0 & 1 & 1 & 0 \\ 0 & 0 & 0 & 0 & 0 & 0 & 1 \\ 0 & 0 & 0 & 0 & 0 & 0 & 0 \\ 0 & 0 & 0 & 0 & 0 & 0 & 0 \\ 0 & 0 & 0 & 0 & 0 & 0 & 0 \end{pmatrix} \end{array},$$

$$\mathcal{M}(\mathbf{G}) = \begin{array}{c} \\ \mathbf{v}_0 \\ \mathbf{v}_1 \\ \mathbf{v}_2 \\ \mathbf{v}_3 \\ \mathbf{v}_4 \\ \mathbf{v}_5 \\ \mathbf{v}_6 \end{array} \begin{array}{c} \begin{array}{cccccc} \mathbf{e}_1 & \mathbf{e}_2 & \mathbf{e}_3 & \mathbf{e}_4 & \mathbf{e}_5 & \mathbf{e}_6 \end{array} \\ \begin{pmatrix} 1 & 0 & 0 & 0 & 0 & 0 \\ -1 & 1 & 1 & 0 & 0 & 0 \\ 0 & -1 & 0 & 1 & 1 & 0 \\ 0 & 0 & -1 & 0 & 0 & 1 \\ 0 & 0 & 0 & -1 & 0 & 0 \\ 0 & 0 & 0 & 0 & -1 & 0 \\ 0 & 0 & 0 & 0 & 0 & -1 \end{pmatrix} \end{array}.$$

It is not difficult to see that the structure of a labeled graph/digraph can be described by its adjacency or incidence matrix. The incidence matrix, in general, is not a square matrix so that it lacks some of the power of the adjacency matrix in computation aspect. Despite this shortcoming, the incidence matrix has some valuable uses. For example, the incidence relationship among all vertices and edges of a graph can be fully described by such a vertex-edge incidence matrix.

If **G** is a labeled digraph derived from a labeled graph G (as shown in Fig. 3.4), there is a simple connection between the adjacency matrix of G, $\mathcal{B}(G)$, and the incidence matrix of **G**, $\mathcal{M}(\mathbf{G})$, such that:

$$\mathcal{B}(G) = \mathcal{D}(\mathbf{G}) - \mathcal{M}(\mathbf{G})\mathcal{M}(\mathbf{G})^T, \tag{3.1}$$

where $\mathcal{D}(\mathbf{G})$ is the $n \times n$ diagonal matrix in which the entry in row i and column i is equal to the degrees of vertex $\mathrm{v_i}$ in **G**, i.e., the number of non-zero entries in the ith row of $\mathcal{M}(\mathbf{G})$.

Definition 3.11 (*Accessibility Matrix*) Let $\mathbf{G} = (V, E)$ be a labeled digraph, where $V = \{\mathrm{v_1, v_2, \ldots, v_m}\}$ and $E = \{\mathrm{e_1, e_2, \ldots, e_n}\}$. The accessibility matrix $\mathcal{R}(\mathbf{G})$ is the $m \times m$ matrix in which the entry in row i and column j is 1, if there is at least one path from $\mathrm{v_i}$ to $\mathrm{v_j}$; it is 0, otherwise.

Example 3.4 The accessibility matrix of the directed graph **G** as shown in Fig. 3.4b can be written as

$$\mathcal{R}(\mathbf{G}) = \begin{array}{c} \\ \mathbf{v}_0 \\ \mathbf{v}_1 \\ \mathbf{v}_2 \\ \mathbf{v}_3 \\ \mathbf{v}_4 \\ \mathbf{v}_5 \\ \mathbf{v}_6 \end{array} \begin{array}{c} \begin{array}{ccccccc} \mathbf{v}_0 & \mathbf{v}_1 & \mathbf{v}_2 & \mathbf{v}_3 & \mathbf{v}_4 & \mathbf{v}_5 & \mathbf{v}_6 \end{array} \\ \begin{pmatrix} 0 & 1 & 1 & 1 & 1 & 1 & 1 \\ 0 & 0 & 1 & 1 & 1 & 1 & 1 \\ 0 & 0 & 0 & 0 & 1 & 1 & 0 \\ 0 & 0 & 0 & 0 & 0 & 0 & 1 \\ 0 & 0 & 0 & 0 & 0 & 0 & 0 \\ 0 & 0 & 0 & 0 & 0 & 0 & 0 \\ 0 & 0 & 0 & 0 & 0 & 0 & 0 \end{pmatrix} \end{array}.$$

The accessibility matrix $\mathcal{R}(\mathbf{G})$ can be simply derived from the powers of the adjacency matrix $\mathcal{B}(\mathbf{G})$ according to the following computation procedure:

1. Let $Q(\mathbf{G}) = \mathcal{B}(\mathbf{G}) + \mathcal{B}^2(\mathbf{G}) + \cdots + \mathcal{B}^m(\mathbf{G})$;
2. Convert $Q(\mathbf{G})$ into $\mathcal{R}(\mathbf{G})$ such that if $q_{ij} \neq 0$, then $r_{ij} = 1$; it is 0, otherwise.

Definition 3.12 (*Path Matrix*) Let $\mathbf{G} = (V, E)$ be a directed tree with m pendant vertices, where $V = \{\mathrm{v_0, v_1, \ldots, v_n}\}$ and $E = \{\mathrm{e_1, e_2, \ldots, e_n}\}$. The path matrix $\mathcal{P}(\mathbf{G})$ is the $m \times (n+1)$ matrix in which the entry in row i and column j is 1, if path i contains $\mathrm{v_{j-1}}$ ($i = 1, 2, \ldots, m$; $j = 1, 2, \ldots, n+1$); it is 0, otherwise.

Example 3.5 The path matrix of the directed (rooted) tree shown in Fig. 3.4b can be written as

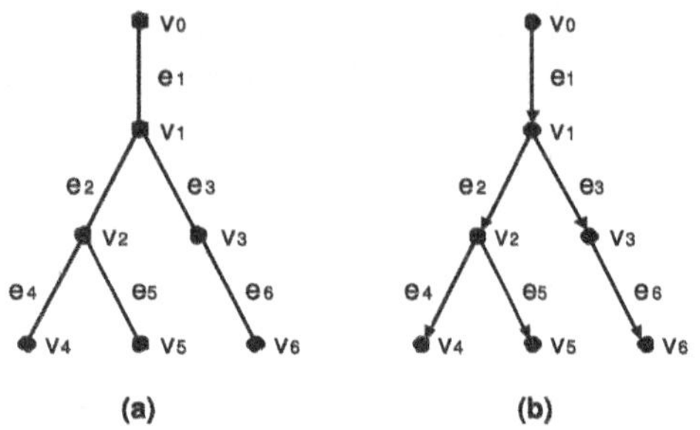

Fig. 3.4 A tree and a directed tree

$$
\mathcal{P}(\mathbf{G}) = \begin{matrix} \\ \mathbf{p}_1 \\ \mathbf{p}_2 \\ \mathbf{p}_3 \end{matrix} \begin{matrix} \begin{matrix} \mathbf{v}_0 & \mathbf{v}_1 & \mathbf{v}_2 & \mathbf{v}_3 & \mathbf{v}_4 & \mathbf{v}_5 & \mathbf{v}_6 \end{matrix} \\ \begin{pmatrix} 1 & 1 & 1 & 0 & 1 & 0 & 0 \\ 1 & 1 & 1 & 0 & 0 & 1 & 0 \\ 1 & 1 & 0 & 1 & 0 & 0 & 1 \end{pmatrix} \end{matrix}.
$$

The path matrix of a directed (rooted) tree $\mathcal{P}(\mathbf{G})$ can be derived from its accessibility matrix $\mathcal{R}(\mathbf{G})$ according to the following computation scheme:

1. Let $\mathcal{U}(\mathbf{G}) = \mathcal{R}(\mathbf{G}) + I_{n \times n}$, where $I_{n \times n}$ is a $n \times n$ identity matrix in which n is the number of vertices;
2. Delete column i of $\mathcal{U}(\mathbf{G})$, if the entries in row i of $\mathcal{R}(\mathbf{G})$ are not all equal to 0;
3. $\mathcal{P}(\mathbf{G})$ is given by the transpose of the resulting $\mathcal{U}(\mathbf{G})$.

A tree G becomes a rooted or directed tree $\mathbf{G}$ when the root vertex is specified. If the incidence matrix of an underlying tree, $\mathcal{M}(G)$, and the root vertex are given, we can derive various matrix representations of $\mathbf{G}$, such as $\mathcal{M}(\mathbf{G})$, $\mathcal{B}(\mathbf{G})$, $\mathcal{R}(\mathbf{G})$, and $\mathcal{P}(\mathbf{G})$, by employing a tree traversing algorithm, e.g., Breadth-First Search (BFS) algorithm or Depth-First Search (DFS) [50, 51] algorithm, to search $\mathcal{M}(G)$. Alternatively, if the rows (vertices) in $\mathcal{M}(\mathbf{G})$ are arranged in the BFS searching sequence as shown in Fig. 3.4a and Example 3.2 (refer to [50] for more details), then $\mathcal{M}(\mathbf{G})$, $\mathcal{B}(\mathbf{G})$, $\mathcal{R}(\mathbf{G})$, and $\mathcal{P}(\mathbf{G})$ can be derived by the following computation procedure:

1. Initialization: Given $\mathcal{M}(G)$;
2. Set $\mathcal{M}(\mathbf{G})$ equal to $\mathcal{M}(G)$ and then let the first row remain unchanged and the first non-zero entry in each of the remaining rows (from left to right) become −1;
3. Set the entry in row i and column i of $\mathcal{D}(\mathbf{G})$ to be equal to the number of non-zero entries in the ith row of $\mathcal{M}(G)$ or $\mathcal{M}(\mathbf{G})$;
4. Compute $\mathcal{B}(G)$ by Eq. (3.1);
5. Set $\mathcal{B}(\mathbf{G})$ equal to $\mathcal{B}(G)$ and then let all the entries in the lower triangular block become zero;
6. Compute $\mathcal{R}(\mathbf{G})$ according to the powers of $\mathcal{B}(\mathbf{G})$;
7. Compute $\mathcal{P}(\mathbf{G})$ from $\mathcal{R}(\mathbf{G})$ by the algorithm proposed previously.

A specialized graph mentioned in the previous section can also be described in an incidence-matrix-like structure as well. An additional row and column are required for the assignments of vertex types and edge types, respectively. Let $\tilde{G}$ be a specialized graph with m vertices and n edges, the following matrix representation is defined for $\tilde{G}$.

Definition 3.13 (*Extended Incidence Matrix* [14]) The Extended Incidence Matrix (eIM) of the specialized graph $\tilde{G}$ denoted by $M(\tilde{G})$ is an $(m+1) \times (n+1)$ matrix such that:

- $m_{ij} = 1$, if $\mathrm{e_j}$ is incident on $\mathrm{v_i}$; $m_{ij} = 0$, otherwise $(i = 1, \ldots, m;\ j = 1, \ldots, n)$;
- $m_{i,n+1} = f_v(\mathrm{v_i}) \in \tilde{V}$, which is $\mathrm{v_i}$'s vertex assignment $(i = 1, \ldots, m)$;
- $m_{m+1,j} = f_e(\mathrm{e_j}) \in \tilde{E}$, which is $\mathrm{e_j}$'s edge assignment $(j = 1, \ldots, n)$;
- $m_{m+1,n+1} = 0$.

The upper-left submatrix of $M(\tilde{G})$ is identical to the incidence matrix of its underlying labeled graph G because they have the identical graph topology. The vertex and edge assignments are kept in the last column and the last row of $M(\tilde{G})$, respectively.

Example 3.6 The eIM of the specialized graph shown in Fig. 3.3b is given by

$$M(\tilde{G}) = \begin{array}{c} \\ \mathbf{v}_1 \\ \mathbf{v}_2 \\ \mathbf{v}_3 \\ \mathbf{v}_4 \\ \mathbf{v}_5 \\ \mathbf{v}_6 \\ \tilde{\mathbf{E}} \end{array} \begin{array}{c} \begin{array}{cccccc} \mathbf{e}_1 & \mathbf{e}_2 & \mathbf{e}_3 & \mathbf{e}_4 & \mathbf{e}_5 & \tilde{\mathbf{V}} \end{array} \\ \begin{pmatrix} 1 & 0 & 0 & 0 & 0 & A \\ 0 & 1 & 0 & 0 & 0 & A \\ 1 & 1 & 1 & 0 & 0 & B \\ 0 & 0 & 1 & 1 & 1 & B \\ 0 & 0 & 0 & 1 & 0 & C \\ 0 & 0 & 0 & 0 & 1 & C \\ \mathrm{x} & \mathrm{x} & \mathrm{y} & \mathrm{z} & \mathrm{z} & 0 \end{pmatrix} \end{array}.$$

3.2 Kinematic Graphs

In mechanism design, kinematic chains of links and joints are often represented by graphs. As a joint can be connected to two links only and a link may accept more than one joint, one logical way to convert a kinematic chain to a graph is to replace the joints with edges and links with vertices. Such a graph representation is called a kinematic graph [52] and has been widely used in mechanism synthesis and design [14, 53–55]. Figure 3.5 shows a planar 6-bar linkage and its kinematic graph. Note that the ground is treated as a fixed link here.

A modular robot is a collection of link and joint modules so that it is essentially a kinematic chain of links and joints. It follows that a modular robot after conversion can also be represented by a kinematic graph. This graph technique provides a framework for the study of underlying geometric and kinematic properties of a modular

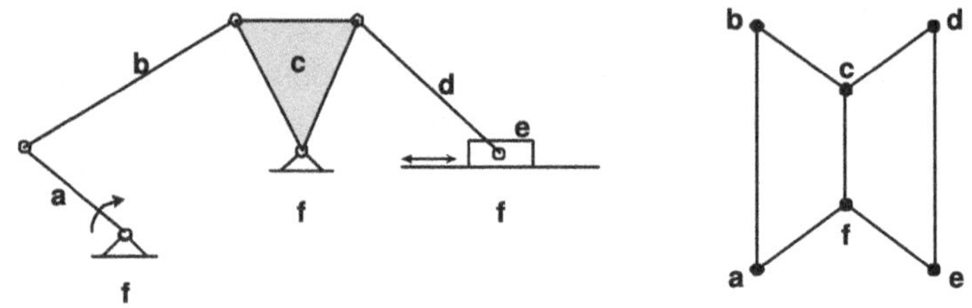

Fig. 3.5 A planar 6-bar linkage and its kinematic graph

robot, for example, the functional difference between a star type graph topology (a walking robot) and a serial type graph topology (an industrial robot). Section 3.4 will show that a modular robot assembly configuration can be fully described by adding more structural and connecting information of modules to the incidence matrix of a kinematic graph. The graph theory and numerous graph algorithms thus can be directly employed for enumeration and classification of modular robot assembly configurations [14].

3.3 Re-classification of Links and Joints

Based on the module designs, any two modules in the Modular Reconfigurable Robot System (MRRS), regardless of their types and sizes, can be connected through a connector. For instance, two joint modules can also be assembled together. Therefore, we cannot directly convert such a modular robot into a kinematic graph just by replacing the joint modules with edges and link modules with vertices. To solve this problem, a re-classification scheme for links and joints is proposed to translate a modular robot into a common kinematic chain.

In this re-classification scheme, each module, regardless of its type, is considered as a "link"; each connector is considered as a "joint" between two consecutive modules. The connecting points on the modules are termed connecting sockets or connecting ports. The connector is treated as a "revolute joint" when fastened to the rotating socket of the revolute joint module; a "prismatic joint" when fastened to the translating socket of the prismatic joint module; and a "fixed joint" when fastened to two fixed sockets of any modules. The fixed joint does not allow any joint displacement, i.e., the two modules are rigidly connected together. Note that, two moving sockets of different modules cannot be connected through a connector in the MRRS.

According to the dimensions of the actually designed modules and the re-classification scheme, the link set, L, consists of six elements: $L=\{L_{c1}, L_{c2}, L_{r1}, L_{r2}, L_{p1}, L_{p2}\}$. L_{c1} and L_{c2} represent the large and the small cubic links (or cubic link modules), respectively; L_{r1} and L_{r2} represent the large and the small revolute links (or revolute link modules), respectively; and L_{p1} and L_{p2} represent the large and the small prismatic links (or prismatic link modules), respectively. Similarly, the joint

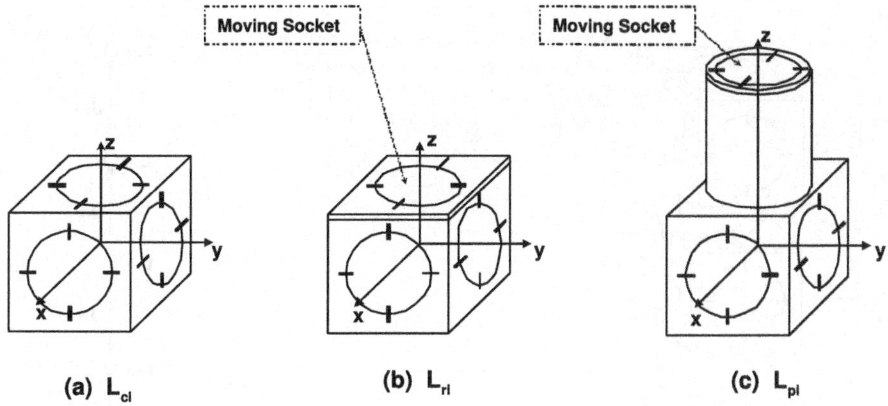

Fig. 3.6 Redefined links

set, J contains nine elements: $J = \{J_{r1}, J_{r2}, J_{r3}, J_{p1}, J_{p2}, J_{p3}, J_{f1}, J_{f2}, J_{f3}\}$, where J_{ri}, J_{pi}, and J_{fi} represent the revolute joint, the prismatic joint, and the fixed joint, respectively. The subscript i indicates the size of a joint, i.e., large ($i = 1$), medium ($i = 2$), and small ($i = 3$) joints. Large, medium, and small joints are derived from the large connector, the adapter, and the small connector, respectively. Note that, hereafter, the term link and joint will indicate the redefined ones if not specified. Based on the actual module designs, the redefined links are shown in Fig. 3.6. For our modeling purpose, each link is attached with a body-fixed frame (or the module frame). The origin of this frame is located at the centroid of the cubic part of a link. The coordinate axes are perpendicular to the faces of modules. If the link is L_{ri} or L_{pi} ($i = 1, 2$), for convenience, the z-axis of the link frame is defined to be along the direction of moving sockets.

After this re-classification, a modular robot structure can be transformed into a common kinematic chain, and thus can be converted into a kinematic graph.

Example 3.7 Figure 3.7a shows the assembly configuration of a modular robot. Its kinematic graph is a specialized graph, as shown in Fig. 3.7b. The eIM of this graph is given by

$$M(\tilde{G}) = \begin{bmatrix} 1 & 0 & 0 & 0 & L_{c1} \\ 1 & 1 & 0 & 0 & L_{r1} \\ 0 & 1 & 1 & 0 & L_{r1} \\ 0 & 0 & 1 & 1 & L_{p2} \\ 0 & 0 & 0 & 1 & L_{c2} \\ J_{f1} & J_{r1} & J_{r2} & J_{p3} & 0 \end{bmatrix}.$$

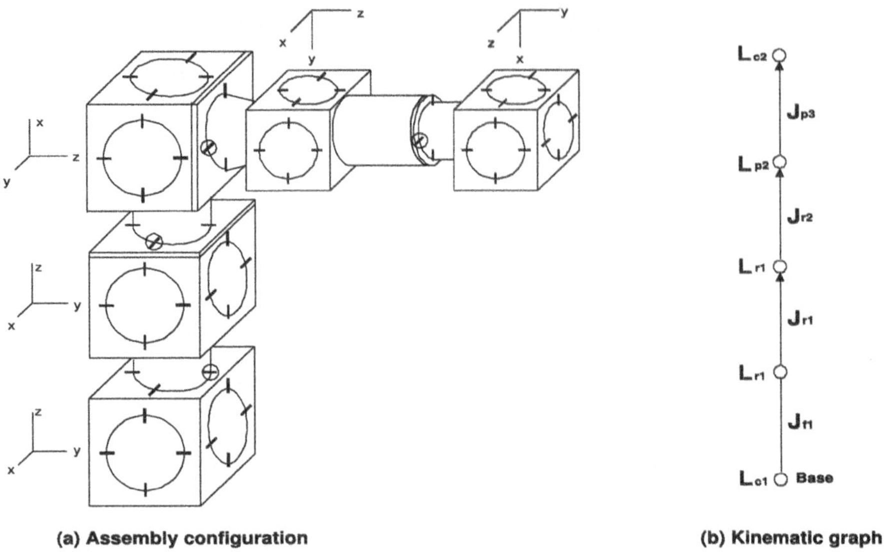

Fig. 3.7 A modular robot assembly configuration

3.4 Assembly Incidence Matrix

A modular robot usually consists of modules of different types and sizes. Thus, it can be topologically described by a specialized graph, and then represented by an eIM. However, the eIM cannot fully represent the assembly configuration of a modular robot because it does not contain the assembly information such as the locations of the connected sockets with respect to the modules and the relative assembly orientation of connected links. Hence, a modified eIM is proposed in order to represent the assembly configuration of a modular robot in a matrix form. Since each of the 1 entries in the eIM defines the connectivity of a link and a joint, a possible approach to modify the eIM is to replace the 1s with specially defined labels to indicate the assembly information between links and joints.

As mentioned in Chap. 2, a joint is allowed to attach a connecting socket of a link in four different orientations as shown in Fig. 3.8. These four symmetric orientations are marked by the numbers 1, 2, 3, and 4 corresponding to four different positions of the location holes. In other words, when the connecting socket and the position of the locating pin are given, the assembly of a link and a joint is uniquely determined. Therefore, the label to be defined should be capable of indicating any connecting socket and the position of the locating pin on a link. Based on the cubic or cube-like shape of links shown in Fig. 3.6, the label is defined by an ordered pair of directional parameters, termed a port vector $P = (p_1, p_2)$, where p_1 represents the normal direction of the connecting port on a link that the connector is engaged, and p_2 describes the position of the locating pin, i.e., the approaching orientation of the connected joint. The directional parameters p_1 and p_2 are all with respect to the link

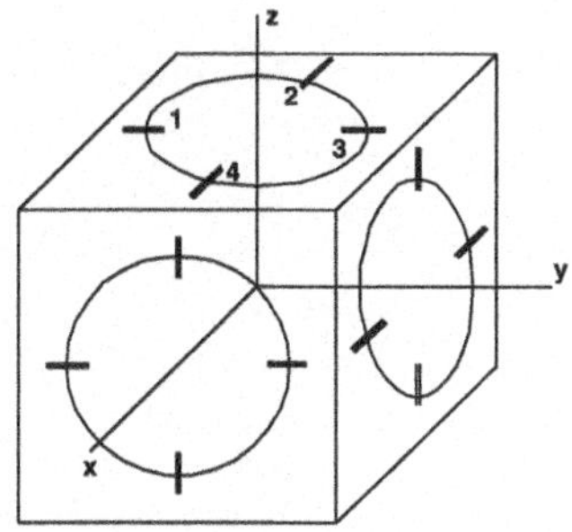

Fig. 3.8 Relative orientations of a joint attaching to a link's connecting port

module frame so that $p_1, p_2 \in \{\pm x, \pm y, \pm z\}$, where $\pm x$, $\pm y$, and $\pm z$ represent the positive and negative directions of x, y, and z axes of the link frame, respectively and $p_1 \neq p_2$. By taking advantage of the inherent relations of the connecting sockets, positions of the location pin, and link frame, the defined labels can conveniently describe the assembly information. For instance, if the connecting socket is on the top surface of the module and the location pin occurs at position "1" as shown in Fig. 3.8, the corresponding port vector for this assembly pattern is given by $P = (+z, -y)$. Note that the representation and assignment of the labels are non-unique. The labels of the connecting sockets can also be represented by other symbols such as numbers [14].

Definition 3.14 (*Assembly Incidence Matrix* [4, 14]) Let $\tilde{G}$ be the specialized graph of a modular robot and $M(\tilde{G})$ be its eIM. The AIM of this robot $\mathcal{A}(\tilde{G})$ is defined by substituting each of 1s in $M(\tilde{G})$ with the corresponding port vector $P = (p_1, p_2)$, where $p_1, p_2 \in \{\pm x, \pm y, \pm z\}$ and $p_1 \neq p_2$.

Example 3.8 The AIM of the modular robot shown in Fig. 3.7a is given by

$$\mathcal{A}(\tilde{G}) = \begin{bmatrix} (+z,+y) & 0 & 0 & 0 & L_{c1}(Base) \\ (-z,+y) & (+z,+x) & 0 & 0 & L_{r1} \\ 0 & (-x,+y) & (+z,+y) & 0 & L_{r1} \\ 0 & 0 & (-z,+x) & (+z,+x) & L_{p2} \\ 0 & 0 & 0 & (-y,+z) & L_{c2} \\ J_{f1} & J_{r1} & J_{r2} & J_{p3} & 0 \end{bmatrix}.$$

Although a modular robot system can render many possible kinematic structures such as tree and closed loop types, we focus on the tree-structured robots that both serial and star-like robots belong to. Since the base module in a specific robot assembly is always known, we, for convenience, always arrange the rows in an AIM in the BFS order. On the other hand, once such an AIM is given, it is very easy to derive the incidence matrix of its underlying graph by first deleting the last column and last row of the AIM and then setting all the non-zero entries to 1. It follows that both the accessibility and path matrices of a (rooted) tree-structured modular robot can be derived from the given AIM automatically. Note that the AIM can also be used to represent closed-loop modular robot configurations without modification [14].

Chapter 4
Modular Serial Robot Kinematics

Abstract This chapter briefly reviews the significant research results in robot kinematics. Based on the local frame representation of the Product-of-Exponentials (POE) formula—the local POE formula, the dyad kinematics between two consecutive links is introduced. Applying the dyad kinematics recursively on the adjacent links, the forward kinematics of a tree-structured robot can be derived by means of the path matrix. The inverse kinematics algorithm follows a numerical approach on top of the Newton-Raphson iteration method. It can deal with pure position, pure orientation, and hybrid inverse kinematics problems.

4.1 Introduction

Robot kinematics is one of the fundamental issues in robotics. It studies the motion of robots for programming, control, and design purposes. Based on the geometric relationships of links and joints, several effective kinematic modeling methods, such as the Denavit-Hartenberg (D-H) method [56], the screw coordinates method [57], the zero reference position method [59], and the Product-of-Exponentials (POE) formula [60] have been developed for conventional robots. Among them, the D-H representation and the POE formula are two significant milestones in robot kinematics.

The D-H method [56] uses a 4×4 homogeneous transformation matrix to describe the spatial relationship between adjacent links. The forward kinematics problem becomes finding an equivalent 4×4 homogeneous transformation matrix that relates the spatial displacement of the end-effector coordinate frame to the base reference coordinate frame. Based on particularly assigned link frames, the number of geometric parameters in the D-H representation of a robot is minimized such that only four geometric parameters, including the joint angles, are required to formulate the homogeneous transformation matrix between two adjacent links. The D-H method is algorithmically universal in deriving the kinematic equations of a robot. However, the D-H representation results in a set of rigidly defined coordinate frames

G. Yang and I.-M. Chen, *Modular Robots: Theory and Practice*, Research on Intelligent Manufacturing,
https://doi.org/10.1007/978-981-16-5007-9_4

in a robot so that arbitrary assignment of link coordinate frames is inconvenient. Also, the D-H parameters are configuration dependent. The same robot may have different sets of D-H parameters just because of the different initial configurations. For a modular robot, the reconfigurability calls for a configuration independent modeling method which is more flexible and adaptable.

Based on the screw theory, Brockett [60] showed that the kinematic equations for an open kinematic chain containing either revolute joints or prismatic joints can be expressed by a product of matrix exponentials (see Sect. 4.2 for more details). Because of its connection with Lie groups and Lie algebras, the POE formula provides a modern geometric interpretation of the classical screw theory. By drawing upon well-established principles of the classical screw theory and the modern differential geometry, the POE formula not only has the intuitive simplicity and theoretical appeal, but can also be used as a computationally effective tool [60–64]. There are several advantages of using the POE formula in robot kinematics modeling. The first advantage is that the POE formula provides a general and uniform representation of the forward kinematics of a spatial mechanism regardless of the types of joints. The second advantage is that the POE formula embodies all the advantages of the screw coordinates method, and allows a global description of the rigid motion which does not suffer from singularities due to the use of local coordinates. Such singularities are inevitable when one chooses to represent the rotation via Euler angles. The third advantage is that the POE formula is basically a zero reference position method so that it also embodies all the advantages of the zero reference position method. The description of the robot is in terms of the joint axis (twists) directions and locations in the zero reference position. Hence, the local coordinate frames in POE representation can be arbitrarily defined. Due to these advantages, the POE formula proved to be a more convenient and powerful modeling method for a Modular Reconfigurable Robot System (MRRS) in which the kinematics, dynamics, and calibration algorithms are to be independent of robot configurations.

The derivation of the forward kinematics for a conventional robot is quite straightforward once a kinematic modeling method is chosen. For a modular robot, the forward kinematics should be formulated automatically once a specific robot configuration (an Assembly Incidence Matrix (AIM)) is given. Researches in CMU [10, 17] and University of Toronto [3] have proposed techniques, based on the D-H representation, to generate modular robot forward kinematics automatically. The method of [10, 17] used two sets of coordinate frames to describe a modular robot: modular frames for individual modules showing kinematic parameters of links and joints of the Reconfigurable Modular Manipulator System (RMMS), and D-H frames defined with D-H notation. According to a given connecting sequence of modules, the D-H parameters of the entire robot can be obtained automatically by employing a conversion algorithm from the modular frames to D-H frames. However, only revolute joints were considered in the kinematic parameter conversion scheme. The work of [3] generalized this modeling technique to include both revolute joints and prismatic joints. Also, the multiple attachment of joints on one link module was considered by defining input/output frames on every link.

Based on the POE formula, another kinematic modeling approach was proposed by Chen and Yang [27], which can generate the forward kinematics of a tree-structured modular robot from a given AIM automatically. In this technique, only one set of local coordinate frames, termed module frames is required to describe the motion of a modular robot. Such a local frame is naturally assigned to each of the modules as shown in Fig. 3.6. A modular robot assembly is then represented by its intrinsic properties, i.e., the locations of the joint axes and the types of the joints, which can be conveniently derived from an AIM by means of a port conversion scheme. The dyad kinematics that relates to the motion of two consecutive modules under a joint displacement was also proposed. Using dyad kinematics along with a tree-traversing algorithm to search for a given AIM, one can derive the forward kinematics automatically.

The inverse kinematics problem consists of the determination of joint variables corresponding to a given end-effector position and orientation. The solution to this problem is of fundamental importance in order to transform the motion specifications assigned to the end-effector in the task space into the corresponding joint space motions that allow the execution of the desired motion.

There are two main approaches, namely, the analytical method and the numerical method, which have been well addressed in the inverse kinematics literatures. The main advantages of the analytical approach over the numerical one are the derivation of computationally efficient closed-form solutions, where each joint variable is expressed in terms of other known quantities, and all the solutions can be found simultaneously [65, 66, 68, 69]. However, the existence of closed form inverse kinematics solutions depends on the kinematic structure of the robot. For example, a sufficient condition for a 6-DOF serial type robot to have a closed form solution is that the three adjacent joint axes intersect together. Therefore, when a general-purpose program for solving the inverse kinematics problem is needed, the numerical method is the preferred approach for solution. Many numerical algorithms have been developed so far. Most of them are based on multi-dimensional Newton-Raphson iteration scheme or similar techniques such as the modified predictor-corrector algorithm, the conjugate gradient algorithm, and the separation of variable method to provide a solution [75–80, 86]. In [81, 82, 101], the inverse kinematic problem was modified as an optimization problem, and numerical optimization algorithms are thus employed to search for the solutions. Most of these inverse kinematics models were based on the D-H notation, and only fixed-configuration, serial type robots are considered.

Note that, in the past two decades, there has also been an increasing interest in the symbolic computation of the inverse kinematics of a serial, articulated robot, especially, the $6R$ robot. Most of these studies have focused on the derivation of high degree polynomials in the tangent of the half-angle of joint variables, and implementation of efficient and robust algorithms to find all of the roots of the polynomial [70–74, 83–85]. Although this approach is general, only serial robots with revolute joints are considered, and at present it is still slow for on-line computations.

For a modular robot system, one can build numerous robots with distinct geometries out of an inventory of robot modules [14]. Therefore, it is impossible and impractical to derive the kinematics model of every modular robot configuration case by case and store the models as library packages in the computer. Khosla et al. [17], based on the D-H notation, proposed a numerical inverse kinematic algorithm for serial type modular robots. In this chapter, based on the local POE formula, we present a systematic modeling scheme which can generate the inverse kinematic model of a modular robot with tree-structured geometry automatically [16]. In this scheme, we use a path matrix of the underlying robot geometry to assist the construction of the inverse kinematics model. Some variations have also been made on the inverse kinematic model for dealing with the pure position, pure orientation, and hybrid inverse kinematics problems. According to the differential form of the resulting inverse kinematic model, a numerical approach using Newton-Raphson iteration method is employed. Note that the main objective of this scheme, like the work of Featherstone [86], is to increase the generality of the inverse kinematic algorithm, which is very important for modular robot systems.

4.2 Geometric Background and the POE Formula

In this section, the POE formula as well as the geometric background is reviewed. For more details, please refer to [60–62, 87].

4.2.1 Geometric Background

In robot kinematics, it is sufficient to think of $SE(3)$, the Special Euclidean Group of rigid body motions, as consisting of matrices of the form

$$\begin{bmatrix} R & p \\ 0 & 1 \end{bmatrix}, \tag{4.1}$$

where $R \in SO(3)$ is interpreted as a rigid body rotation and $p \in \Re^{3\times 1}$ as a rigid body translation. Here, $SO(3)$, the Special Orthogonal Group, denotes the group of 3×3 rotation matrices. Elements of $SE(3)$ can also be denoted by the ordered pair (p, R), with group multiplication understood to be $(p_1, R_1) \cdot (p_2, R_2) = (R_1 p_2 + p_1, R_1 R_2)$. $SE(3)$ is a Lie group of dimension six.

The Lie algebra of $SE(3)$, denoted $se(3)$, consists of matrices of the form

$$\begin{bmatrix} \hat{w} & v \\ 0 & 0 \end{bmatrix}, \tag{4.2}$$

where

$$\hat{w} = \begin{bmatrix} 0 & -w_z & w_y \\ w_z & 0 & -w_x \\ -w_y & w_x & 0 \end{bmatrix}. \tag{4.3}$$

The set of 3×3 real skew-symmetric matrices, $\hat{w}$, forms the Lie algebra of $SO(3)$, denoted by $so(3)$. Note that an element of $\hat{w} \in so(3)$ can also be regarded as a vector

$w \in \Re^{3\times1}$. Since in most cases it will be clear from the context which representation is implied, an element of $so(3)$, $\hat{w}$, will also be simply denoted by w in such cases. An element of $se(3)$ will thus admit a 6×1 vector presentation: $(v, w) \in \Re^{6\times1}$, termed a twist. The twist denotes the line coordinate of the screw axis of a general rigid body motion (rotation and translation) in which w is the unit directional vector of the axis and v is the position of the axis relative to the origin.

On matrix Lie algebras, the Lie bracket is given by the matrix commutator: if A and B are elements of a matrix Lie algebra, then $[A, B] = AB - BA$. In particular, on $so(3)$ the Lie bracket of two elements corresponds to their vector product: $[w_1, w_2] = w_1 \times w_2$. On $se(3)$, the Lie bracket of two elements (v_1, w_1) and (v_2, w_2) is given by

$$[(v_1, w_1), (v_2, w_2)] = (w_1 \times v_2 - w_2 \times v_1, w_1 \times w_2). \tag{4.4}$$

An element of a Lie group can also be identified with a linear mapping between its Lie algebra via the adjoint representation. Suppose G is a matrix Lie group with Lie algebra g. For every $X \in G$, the adjoint map $Ad_X : g \to g$ is defined by $Ad_X(x) = XxX^{-1}$ for $x \in g$. If $X = (p, R)$ is an element of $SE(3)$, then its adjoint map acting on an element $x = (v, w)$ of $se(3)$ is given by

$$Ad_X(x) = (p \times Rw + Rv, Rw), \tag{4.5}$$

which also admits the 6×6 matrix representation

$$Ad_X(x) = \begin{bmatrix} R & \hat{p}R \\ 0 & R \end{bmatrix} \begin{bmatrix} v \\ w \end{bmatrix}. \tag{4.6}$$

It is easily verified that $Ad_X^{-1} = Ad_{X^{-1}}$ and $Ad_X Ad_Y = Ad_{XY}$ for any $X, Y \in SE(3)$.

Elements of a Lie algebra can also be identified with a linear mapping between its Lie algebra via the Lie bracket. Given an element $x \in g$, its adjoint representation is the linear map $ad_x : g \to g$ defined by $ad_x(y) = [x, y]$. If $x = (v_1, w_1)$ and $y = (v_2, w_2)$ are elements of $se(3)$, then

$$ad_x y = (w_1 \times v_2 - w_2 \times v_1, w_1 \times w_2), \tag{4.7}$$

which also admits the matrix representation

$$ad_x y = \begin{bmatrix} \hat{w}_1 & \hat{v}_1 \\ 0 & \hat{w}_1 \end{bmatrix} \begin{bmatrix} v_2 \\ w_2 \end{bmatrix}. \tag{4.8}$$

An important connection between a Lie group, $SE(3)$, and its Lie algebra, $se(3)$, is the exponential mapping, defined on each Lie algebra. Let $\hat{s} \in se(3)$ $(s = (v, w))$, and $\|w\|^2 = w_x^2 + w_y^2 + w_z^2$, then

$$e^{\hat{s}} = \begin{bmatrix} e^{\hat{w}} & Av \\ 0 & 1 \end{bmatrix} \in SE(3), \tag{4.9}$$

where

$$e^{\hat{w}} = I + \frac{\sin\|w\|}{\|w\|}\hat{w} + \frac{1-\cos\|w\|}{\|w\|^2}\hat{w}^2, \tag{4.10}$$

$$A = I + \frac{1-\cos\|w\|}{\|w\|^2}\hat{w} + \frac{\|w\| - \sin\|w\|}{\|w\|^3}\hat{w}^2. \tag{4.11}$$

The matrix logarithm also establishes a connection between a Lie group, $SE(3)$, and its Lie algebra, $se(3)$ [87]. Let $R \in SO(3)$ such that trace$(R) \neq -1$, $1 + 2\cos\phi =$ trace(R), and $\|\phi\| < \pi$, then

$$\log\begin{bmatrix} R & p \\ 0 & 1 \end{bmatrix} = \begin{bmatrix} \hat{w} & A^* p \\ 0 & 0 \end{bmatrix} \in se(3), \tag{4.12}$$

where

$$\hat{w} = \log R = \frac{\phi}{2\sin\phi}(R - R^T), \tag{4.13}$$

$$A^* = I - \frac{1}{2}\hat{w} + \frac{2\sin\|w\| - \|w\|(1+\cos\|w\|)}{2\|w\|^2\sin\|w\|}\hat{w}^2. \tag{4.14}$$

If ϕ is very small, $\hat{w} \approx (R - R^T)/2$. If the trace$(R) = -1$, it can be shown that $\log R = (2k+1)\pi\hat{v}$, where k is any integer and v is the unit eigenvector of $\log R$ corresponding to the eigenvalue 1.

An actual example of Lie algebra in robot kinematics is the generalized velocities of rigid bodies [62, 87]. There are two ways in which the tangent vector $\dot{X}(t)$ of a curve $X(t) = (p(t), R(t)) \in SE(3)$ can be identified with an element of $se(3)$. If $X(t)$ describes the motion of a rigid body relative to a spatial reference frame, then both $\dot{X}X^{-1}{=}(\dot{p} - \dot{R}R^{-1}p, \dot{R}R^{-1})$ and $X^{-1}\dot{X} = (R^{-1}\dot{p}, R^{-1}\dot{R})$ are elements of $se(3)$. $X^{-1}\dot{X}$ is referred to as the body velocity representation of $\dot{X}$, since $R^{-1}\dot{R}$ and $R^{-1}\dot{p}$ are the angular velocity and translational velocity of the rigid body relative to its body-fixed frame, respectively. Similarly, $\dot{X}X^{-1}$ is referred to as the spatial velocity representation of $\dot{X}$. One subtle and important difference in the interpretation of the spatial velocity is that $R^{-1}\dot{R}$ is indeed the angular velocity of the rigid body relative to the spatial frame, while the translational velocity relative to the spatial frame is not $\dot{p} - \dot{R}R^{-1}p$, but simply $\dot{p}$. If $X(t)$ undergoes a coordinate transformation of the form $X(t) \mapsto X(t)T$, where $T \in SE(3)$ is constant, then its new body velocity representation is $T^{-1}X^{-1}\dot{X}T = Ad_{T^{-1}}(X^{-1}\dot{X})$.

4.2.2 The POE Formula

We now briefly review the POE formula for open chain robots in which the links form a single serial chain and each pair of links is connected either by a revolute joint or a prismatic joint [60, 62, 87]. For convenience, we temporarily number the joints from 1 to n, starting at the base, and number the links such that joint i connects links $i-1$ and i. Link 0 is taken to be the base and link n is taken to be the end link. If a right-handed reference frame is fixed at each link of the chain, then the element of $SE(3)$ describing the position and orientation of frame i relative to frame $i-1$ is $T_{i-1,i}(q_i)= T_{i-1,i}(0)e^{\hat{s}_{i,i}q_i}$, where $T_{i-1,i}(0) \in SE(3)$ is the initial position and orientation of frame i relative to frame $i-1$, $\hat{s}_{i,i}$ or simply $\hat{s}_i \in se(3)$ represents the twist of joint i's axis expressed in frame i, and $q_i \in \Re$ is the joint i's displacement. The frame fixed at the end-effector or the end-link is then related to that of the base by the product

$$
\begin{aligned}
T_{0,n}(q_1, q_2, \ldots, q_n) &= T_{0,1}(q_1)T_{1,2}(q_2)\ldots T_{n-1,n}(q_n) \\
&= T_{0,1}(0)e^{\hat{s}_1 q_1}T_{1,2}(0)e^{\hat{s}_2 q_2}\ldots T_{n-1,n}(0)e^{\hat{s}_n q_n}.
\end{aligned} \tag{4.15}
$$

If the twist of joint i's axis is expressed in frame $i-1$, denoted by $\hat{s}_{i-1,i}$, Eq. (4.15) can be rewritten as

$$
T_{0,n}(q_1, q_2, \ldots, q_n) = e^{\hat{s}_{0,1}q_1}T_{0,1}(0)e^{\hat{s}_{1,2}q_2}T_{1,2}(0)\ldots e^{\hat{s}_{n-1,n}q_n}T_{n-1,n}(0), \tag{4.16}
$$

where $\hat{s}_{i-1,i} = T_{i-1,i}(0)\hat{s}_i T_{i-1,i}^{-1}(0) = Ad_{T_{i-1,i}(0)}s_i$.

If all the twists are expressed in the base frame 0, Eq. (4.15) can also be written as

$$
T_{0,n}(q_1, q_2, \ldots, q_n) = e^{\hat{s}_{0,1}q_1}e^{\hat{s}_{0,2}q_2}\ldots e^{\hat{s}_{0,n}q_n}T_{0,n}(0), \tag{4.17}
$$

where $\hat{s}_{0,i} = T_{0,i}(0)\hat{s}_i T_{0,i}^{-1}(0) = Ad_{T_{0,i}(0)}s_i$, in which $T_{0,i}(0) = T_{0,1}(0)T_{1,2}(0)\ldots T_{i-1,i}(0)$. $T_{0,n}(0) = \prod_{i=1}^{n} T_{i-1,i}(0)$ represents the initial position and orientation of the end-effector frame relative to the base frame.

If all of the twists are expressed in the end-effector (end link) frame n, Eq. (4.15) can also be rewritten as

$$
T_{0,n}(q_1, q_2, \ldots, q_n) = T_{0,n}(0)e^{\hat{s}_{n,1}q_1}e^{\hat{s}_{n,2}q_2}\ldots e^{\hat{s}_{n,n}q_n}, \tag{4.18}
$$

where $\hat{s}_{n,i} = T_{i,n}^{-1}(0)\hat{s}_i T_{i,n}(0) = Ad_{T_{i,n}^{-1}(0)}s_i$, in which $T_{i,n}(0) = T_{i,i+1}(0)T_{i+1,i+2}(0)\ldots T_{n-1,n}(0)$.

Equations (4.15)–(4.18) are all called the POE formula, in which Eqs. (4.15) and (4.16) are called the local frame representation of POE formula, i.e., the local POE formula. The POE formula provides a clear geometric representation of the kinematics of a robot and can simplify the kinematic analysis of the mechanism [88].

4.3 Forward Kinematics

In this section, we focus on the forward kinematics of tree-structured modular robots which most of the industrial robots belong to. A tree-structured kinematic chain generally consists of several branches (from the base to each of the pendent links). The forward kinematics of an n-DOF tree-structured modular robot can thus be described as: given a set of joint angles $q = \{q_1, q_2, \ldots, q_n\}$, we wish to determine the poses of all the end-effector frames relative to the unique base frame. Although different branches may have several common links and joints, in formulating the forward kinematics, each branch can still be considered as a serial chain. Since each joint connects two and only two consecutive links, it is reasonable to first consider the description of kinematic relationship between two consecutive links and then obtain the overall description of robot kinematics in a recursive fashion. Therefore, the local frame representation of the POE formula is more suitable for our purpose. Note that in the following context, we consider modular robots without end-effectors, so that the desired poses are referred to as the poses of end link frames.

4.3.1 Dyad Kinematics

As a first step, a systematic and general method is to be derived in order to define the relative position and orientation of two consecutive links. The problem is to determine two frames attached to the two links and compute the coordinate transformations between them. In general, these two frames can be arbitrarily chosen as long as they are attached to the link they are referring to. Since there exists a local frame (module frame) on each of the links, it is unnecessary and inconvenient to define new frames on the links. For convenience, we adopt these naturally defined module frames as the local frames throughout this book.rr

A dyad is referred to as a pair of consecutive links in a kinematic chain. Based on the local POE formula, i.e., Eq. (4.15), the dyad kinematics which relates the motion of two connected modules under a joint displacement is defined as follows:

Let $\mathrm{v_i}$ and $\mathrm{v_j}$ be two adjacent links connected by a joint $\mathrm{e_j}$, as shown in Fig. 4.1 and also consider joint $\mathrm{e_j}$ and link $\mathrm{v_j}$ as link assembly j. Denote the module coordinate frames on link $\mathrm{v_i}$ as frame i and link $\mathrm{v_j}$ as frame j. Then the relative pose (position and orientation) of frame j with respect to frame i, under a joint displacement, q_j, can be described by a 4×4 homogeneous matrix, an element of $SE(3)$, such that

$$T_{ij}(q_j) = T_{ij}(0)e^{\hat{s}_j q_j}, \tag{4.19}$$

where $\hat{s}_j \in se(3)$ is the twist of joint $\mathrm{e_j}$ expressed in frame j, $T_{ij}(q_j)$ and $T_{ij}(0) \in SE(3)$. Note that in the following context, the pose of a coordinate frame is referred to as the 4×4 homogeneous matrix including both the orientation and position of the frame with respect to a reference frame. In Eq. (4.19), $T_{ij}(0)$ is the initial pose

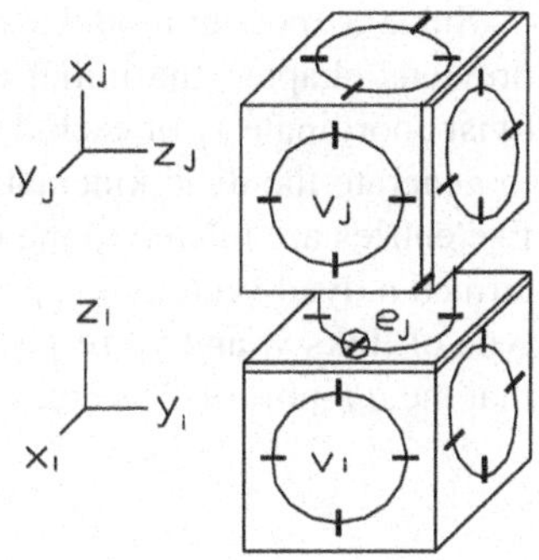

Fig. 4.1 Two connected modules (a dyad)

of frame j relative to frame i, and

$$T_{ij}(0) = \begin{bmatrix} R_{ij}(0) & d_{ij}(0) \\ 0 & 1 \end{bmatrix}, \tag{4.20}$$

where $R_{ij}(0) \in SO(3)$ and $d_{ij}(0) \in \Re^{3\times 1}$ are the initial orientation and position of link frame j relative to frame i, respectively. The twist of link assembly j is written as

$$\hat{s}_j = \begin{bmatrix} \hat{w}_j & v_j \\ 0 & 0 \end{bmatrix}, \tag{4.21}$$

where $\hat{w}_j$ is a skew-symmetric matrix related to $w_j = (w_{jx}, w_{jy}, w_{jz})^T$, the unit directional vector of the joint axis expressed in frame j, and $v_j = (v_{jx}, v_{jy}, v_{jz})^T$ is the position vector of the joint axis relative to frame j. $\hat{w}_j$ is given by

$$\hat{w}_j = \begin{bmatrix} 0 & -w_{jz} & w_{jy} \\ w_{jz} & 0 & -w_{jx} \\ -w_{jy} & w_{jx} & 0 \end{bmatrix}. \tag{4.22}$$

The twist, $\hat{s}_j$, can also be represented by a 6-dimensional vector through a mapping: $\hat{s}_j \mapsto (v_j, w_j) \in \Re^{6\times 1}$, termed twist coordinates. Based on the module design, the joint j's axis always passes through frame j's origin. For revolute joints, $s_j = (0, w_j)$, where w_j is the unit directional vector of the joint axis expressed in frame j. For prismatic joints, $s_j = (v_j, 0)$, where v_j is the unit directional vector of the joint axis relative to frame j. For fixed joints, $s_j = (0, 0)$. Therefore, Eq. (4.9) can also be simplified as

$$e^{\hat{s}_j q_j} = \begin{bmatrix} e^{\hat{w}_j q_j} & v_j q_j \\ 0 & 1 \end{bmatrix}, \tag{4.23}$$

where q_j is the joint j's displacement and

$$e^{\hat{w}_j q_j} = I + \hat{w}_j \sin q_j + \hat{w}_j^2(1 - \cos q_j). \tag{4.24}$$

Since a specific modular robot is given in an AIM format as mentioned in the previous chapter, the initial orientation $R_{ij}(0)$, the initial position $d_{ij}(0)$, and the twist coordinate s_j of each dyad are to be determined from the given AIM in order to generate the dyad kinematics automatically. In an AIM, as shown in Eq. (4.25), five entries are related to the dyad of links v_i and v_j, i.e., $(a_{ik}, a_{in}, a_{jk}, a_{jn}, a_{n+1,k})$, termed a dyad vector. $a_{n+1,k}$ represents the type of joint e_k; a_{in} and a_{jn} indicate the type of links v_i and v_j, respectively; $a_{ik} = \{p_{ik}, q_{ik}\}$ and $a_{jk} = \{p_{jk}, q_{jk}\}$ represent that the a_{ik}-port of v_i is connected with a_{jk}-port of v_j by a joint e_k.

$$\mathcal{A}(\widetilde{G}) = \begin{bmatrix} \cdots & \cdots & \cdots\cdots \\ \cdots & a_{ik} & \cdots\ a_{in} \\ \cdots & \cdots & \cdots\cdots \\ \cdots & a_{jk} & \cdots\ a_{jn} \\ \cdots & \cdots & \cdots\cdots \\ \cdots & a_{n+1,k} & \cdots\ \ 0 \end{bmatrix}. \tag{4.25}$$

- **Determination of Initial Orientation $R_{ij}(0)$**

The columns of $R_{ij}(0)$ represent the initial orientation of the axes of frame j relative to frame i. Furthermore, the first, second, and third columns of $R_{ij}(0)$, denoted by $R_{ij}^m(0)(m = 1, 2, 3)$, represent the unit directional vector of x, y, and z-axis of frame j relative to frame i initially. Let $P_{ik} = a_{ik} = (p_{ik}, q_{ik})$ and $P_{jk} = a_{jk} = (p_{jk}, q_{jk})$ be the port vectors of the two connected modules i and j, respectively. Since these two connecting sockets are facing each other, the direction of p_{jk} of frame j is opposite to the direction of the p_{ik} of frame i. According to this relationship, the direction vector of one axis of frame j (related to p_{jk}) with respect to frame i can be determined, which is opposite to the direction of p_{ik} and then put in the corresponding column of $R_{ij}(0)$ according to p_{jk}. For instance, in Fig. 4.2, if $P_{ik} = (p_{ik}, q_{ik}) = (+z_i, +x_i)$ and $P_{jk} = (p_{jk}, q_{jk}) = (-x_j, +y_j)$, then the direction of x-axis of frame j, x_j, is now in the z-direction of frame i, $+z_i$, so that the vector (0, 0, 1) should be put into the first column of $R_{ij}(0)$.
In the port vectors, q_{ik} and q_{jk} represent the attaching orientations of joint e_k on links v_i and v_j, respectively, which are corresponding to the positions of the location pins. They are in the same physical direction according to our definition. Hence, the direction of another axis of frame j (related to q_{jk}) with respect to frame i can also be determined. It is in the direction of q_{ik} and then put into the corresponding column of $R_{ij}(0)$ according to q_{jk}. Following the previous example (Fig. 4.2), since the y direction of frame j, y_j, is in the x direction of frame i, the unit vector (1, 0, 0) is placed in the second column of $R_{ij}(0)$.
According to the port vectors of two connected links, we can get two column vectors of the initial orientation matrix $R_{ij}(0)$. Since $R_{ij} \in SO(3)$ is an orthogonal matrix, the remaining column vector can be calculated by the cross product of the two known column vectors using one of the following equations:

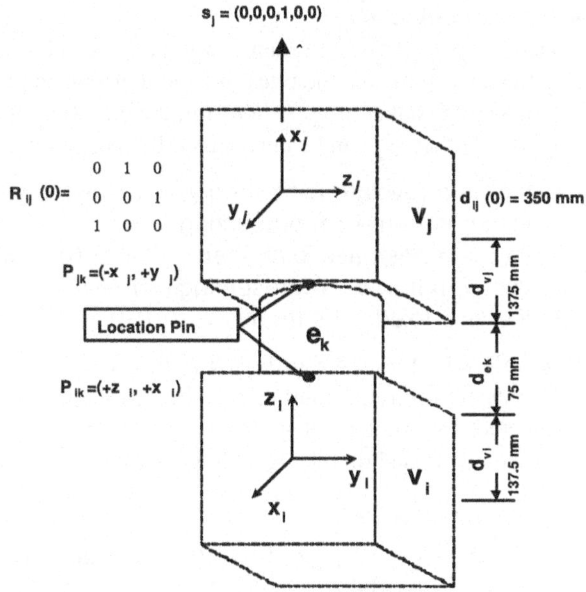

Fig. 4.2 Determination of the initial pose of a dyad

$$R^1_{ij}(0) = R^2_{ij}(0) \times R^3_{ij}(0), \tag{4.26}$$

$$R^2_{ij}(0) = R^3_{ij}(0) \times R^1_{ij}(0), \tag{4.27}$$

$$R^3_{ij}(0) = R^1_{ij}(0) \times R^2_{ij}(0). \tag{4.28}$$

- **Determination of Initial Displacement** $d_{ij}(0)$
  The initial displacement vector represents the position of the origin of frame j relative to frame i initially. $d_{ij}(0) = (d^x_{ij}, d^y_{ij}, d^z_{ij}) \in \Re^{3\times 1}$, where d^x_{ij}, d^y_{ij}, and d^z_{ij} represent the x, y, and z coordinates of frame j's origin relative to frame i, respectively. Because all the connecting sockets of a module are centered on the faces of the cube and the module frame is assigned to the centroid of the cube, the origin of frame j is always in one of frame i's axes, i.e., along the direction of p_{ik}. Therefore, two elements in this vector must be zero. According to p_{ik}, we can first determine which element in $d_{ij}(0)$ is the non-zero entry, and also the entry's sign. The absolute value of the non-zero entry is equal to the distance between the origin of frame i and the origin of frame j, which can be calculated by

$$||d_{ij}(0)|| = d_{\mathrm{vi}} + d_{\mathrm{vj}} + d_{\mathrm{ek}}, \tag{4.29}$$

where d_{vi} (or d_{vj}) is the distance from the origin of link frame i (or j) to the connecting surface of a_{ik}-port (or a_{jk}-port), and d_{ek} is the length of joint (connector) $\mathrm{e_k}$. All these dimensions can be derived from the module database according to a given dyad vector.

- **Determination of** s_j
  Here, s_j is the twist coordinate of joint e_k, which is expressed in frame j. Since v_i is the ancestor of v_j, the twist $\hat{s}_j$ should, according to our assembly rules, point to connecting socket of v_j. As a result, the direction of $\hat{s}_j$ is opposite to the direction of p_{jk}. Hence, s_j can be determined by the following procedure:

1. If e_k is a revolute joint, then $s_j = (0, w_j)^T$, where w_j is a unit directional vector in frame j representing $(-p_{jk})$.
2. If e_k is a prismatic joint, then $s_j = (\upsilon_j, 0)^T$, where υ_j is a unit directional vector in frame j representing $(-p_{jk})$.
3. If e_k is a fixed joint, then $s_j = (0, 0)^T$.

Example 4.1 As shown in Fig. 4.2, link v_i of type L_{r1} and link v_j of type L_{r1} are connected through joint e_k of type J_{r1}. Link v_i is the ancestor of Link v_j. The connected port vectors of v_i and v_j are $P_{ik} = (+z_i, +y_i)$ and $P_{jk} = (-x_j, +y_j)$, respectively. Based on the algorithms mentioned above, we can obtain:

$$R_{ij}(0) = \begin{bmatrix} 0 & 1 & 0 \\ 0 & 0 & 1 \\ 1 & 0 & 0 \end{bmatrix}, \; d_{ij}(0) = (0, 0, 350 \text{ mm}), \text{ and } s_j = (0, 0, 0, 1, 0, 0)^T.$$

Then the forward kinematics transformation from frame i to j can be calculated by Eq. (4.19).

4.3.2 *Forward Kinematics for a Tree-Structured Modular Robot*

The forward kinematics of a tree-structured robot is to determine the kinematic transformations from the base to each of the pendant links (or the end-effectors). Hence, finding out the connecting orders of links from a given AIM is essential. There are two tree-traversing algorithms usually used in graphs, the Breadth-First-Search (BFS) algorithm and the Depth-First-Search (DFS) algorithm. They can both be employed to determine the connecting orders of links as well as the corresponding dyad vectors from a given AIM.

A recursive forward kinematics algorithm for a tree-structured modular robot was proposed by Chen and Yang [27] by using the BFS algorithm to traverse a given AIM and applying Eq. (4.19) to every dyad. The input of this algorithm is a given AIM and a set of joint angles; the output is the poses of all the end links with respect to the base frame. Alternatively, the forward kinematic equation can be derived by means of the path matrix. Since the rows in the AIM are arranged in BFS order, the path matrix can be derived from the given AIM as mentioned in Chap. 3. Once the path matrix is obtained, the dyad vectors can be derived from the AIM accordingly.

Now we consider a tree-structured modular robot consisting of $n + 1$ modules and having m branches. Therefore, the path matrix of this robot has m rows and

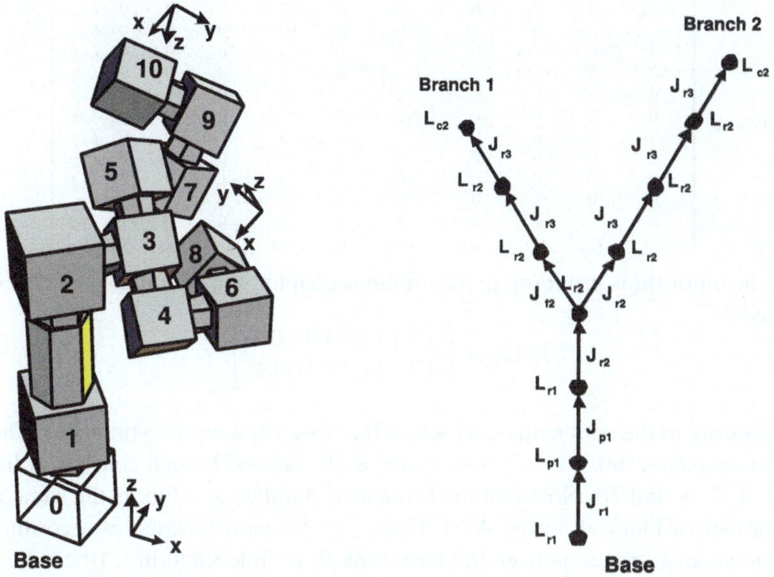

Fig. 4.3 A tree-structured modular robot configuration (9-DOF)

$n + 1$ columns. Suppose that row k ($k = 1, 2, \ldots, m$) has $n[k] + 1$ non-zero entries, sequentially 0_k (or simplify to be 0 because of the unique base), $1_k, \ldots, n[k]_k$ from the left to right. Here i_k is the index of the ith non-zero entry in row k, and $i = 0, 1, \ldots, n[k]$. The forward kinematics of branch k thus can be given by:

$$\begin{aligned} T_{0,n[k]_k} &= T_{0,1_k}(q_{1_k})T_{1_k,2_k}(q_{2_k})\ldots T_{(n[k]-1)_k,n[k]_k}(q_{n[k]_k}) \\ &= \prod_{i=1}^{n[k]}(T_{(i-1)_k,i_k}(0)e^{\hat{s}_{i_k}q_{i_k}}). \end{aligned} \tag{4.30}$$

The forward kinematic equation for the tree-structured modular robot with m branches is

$$T(q_1, q_2, \ldots, q_n) = \begin{bmatrix} T_{0,n[1]_1} \\ T_{0,n[2]_2} \\ \ldots \\ T_{0,n[m]_m} \end{bmatrix} = \begin{bmatrix} \prod_{i=1}^{n[1]}(T_{(i-1)_1,i_1}(0)e^{\hat{s}_{i_1}q_{i_1}}) \\ \prod_{i=1}^{n[2]}(T_{(i-1)_2,i_2}(0)e^{\hat{s}_{i_2}q_{i_2}}) \\ \ldots \\ \prod_{i=1}^{n[m]}(T_{(i-1)_m,i_m}(0)e^{\hat{s}_{i_m}q_{i_m}}) \end{bmatrix}. \tag{4.31}$$

Example 4.2 The AIM of a tree-structured modular robot (as shown in Fig. 4.3) is given by

$$\mathcal{A}(\tilde{G}) = \begin{bmatrix} (z,x) & 0 & 0 & 0 & 0 & 0 & 0 & 0 & 0 & 0 & L_{r1} \\ (-z,y) & (z,x) & 0 & 0 & 0 & 0 & 0 & 0 & 0 & 0 & L_{p1} \\ 0 & (-x,y) & (z,x) & 0 & 0 & 0 & 0 & 0 & 0 & 0 & L_{r1} \\ 0 & 0 & (x,y) & (-z,-y) & (z,y) & 0 & 0 & 0 & 0 & 0 & L_{r2} \\ 0 & 0 & 0 & (-x,y) & 0 & (z,y) & 0 & 0 & 0 & 0 & L_{r2} \\ 0 & 0 & 0 & 0 & (-y,-x) & 0 & (z,x) & 0 & 0 & 0 & L_{r2} \\ 0 & 0 & 0 & 0 & 0 & (y,z) & 0 & (z,x) & 0 & 0 & L_{r2} \\ 0 & 0 & 0 & 0 & 0 & 0 & (-x,y) & 0 & (z,y) & 0 & L_{r2} \\ 0 & 0 & 0 & 0 & 0 & 0 & 0 & (-z,x) & 0 & 0 & L_{c2} \\ 0 & 0 & 0 & 0 & 0 & 0 & 0 & 0 & (x,z) & (z,y) & L_{r2} \\ 0 & 0 & 0 & 0 & 0 & 0 & 0 & 0 & 0 & (y,z) & L_{c2} \\ J_{r1} & J_{p1} & J_{r2} & J_{f2} & J_{r2} & J_{r3} & J_{r3} & J_{r3} & J_{r3} & J_{r3} & 0 \end{bmatrix}.$$

Using the algorithms proposed in the previous chapter, the path matrix of $\mathcal{A}(\tilde{G})$ can be derived:

$$\mathcal{P}(\mathbf{G}) = \begin{bmatrix} 1\,1\,1\,1\,1\,0\,1\,0\,1\,0\,0 \\ 1\,1\,1\,1\,0\,1\,0\,1\,0\,1\,1 \end{bmatrix}.$$

According to the path matrix, we know that this robot has two branches. The first branch consists of links 0, 1, 2, 3, 4, 6, and 8; the second branch consists of links 0, 1, 2, 3, 5, 7, 9, and 10. Note that the sequential number of a link is the same as the row number of that link in the AIM. Using our forward kinematics algorithm, the kinematic transformations from the base (link 0) to link 8 and link 10 can be given by:

$$T(q_1, q_2, \ldots, q_{10}) = \begin{bmatrix} T_{0,8} \\ T_{0,10} \end{bmatrix}$$

$$= \begin{bmatrix} \left(\prod_{i=1}^{3} T_{i-1,i}(0)e^{\hat{s}_i q_i}\right) T_{3,4}(0)e^{\hat{s}_4 q_4} T_{4,6}(0)e^{\hat{s}_6 q_6} T_{6,8}(0)e^{\hat{s}_8 q_8} \\ \left(\prod_{i=1}^{3} T_{i-1,i}(0)e^{\hat{s}_i q_i}\right) T_{3,5}(0)e^{\hat{s}_5 q_5} T_{5,7}(0)e^{\hat{s}_7 q_7} T_{7,9}(0)e^{\hat{s}_9 q_9} T_{9,10}(0)e^{\hat{s}_{10} q_{10}} \end{bmatrix}.$$

Figure 4.3 shows the robot configuration with joint angles $q = [\frac{\pi}{4}, 100 \text{ mm}, \frac{\pi}{4}, 0, \frac{\pi}{4}, \frac{\pi}{4}, \frac{\pi}{4}, \frac{\pi}{4}, \frac{\pi}{4}, \frac{\pi}{4}]^T$.

4.4 Inverse Kinematics

The purpose of an inverse kinematics algorithm is to determine the joint angles that cause a robot to reach a desired pose of the end-effector (the end link). Because a modular robot has unfixed assembly configurations and the number of DOFs in the robot varies, it is usually impossible to find the closed-form inverse kinematics solution for a modular robot. Hence, a numerical approach is employed. In general, a numerical inverse kinematics model can be derived through linearizing the forward kinematic equation. Based on the local POE formula and the definition of logarithm on $SE(3)$ and $SO(3)$, a compact and computational effective inverse kinematics model for a general tree-structured robot was derived by Chen and Yang [16]. This model can be easily modified to solve the pure position, pure orientation, and hybrid inverse kinematics problems. This section starts from formulating the inverse kinematics model of a single branch of a tree-structured robot. The obtained formula is suitable for serial and star-like robots which are the commonly used robot geometries. Then the inverse kinematics model as well as a closed-loop iterative algorithm is proposed for a complete tree-structured robot.

4.4.1 Differential Kinematics Model for a Single Branch

According to Eq. (4.30), the linearized kinematic equation for branch k can be given by

$$dT_{0,n[k]_k} = \sum_{i=1}^{n[k]} \frac{\partial T_{0,n[k]_k}}{\partial q_{i_k}} dq_{i_k}. \tag{4.32}$$

Since

$$\begin{aligned} \frac{\partial T_{0,n[k]_k}}{\partial q_{i_k}} &= T_{0,(i-1)_k} \frac{\partial (T_{(i-1)_k,i_k}(0) e^{\hat{s}_{i_k} q_{i_k}})}{\partial q_{i_k}} T_{i_k,n[k]_k} \\ &= T_{0,(i-1)_k} T_{(i-1)_k,i_k}(0) e^{\hat{s}_{i_k} q_{i_k}} \hat{s}_{i_k} T_{i_k,n[k]_k} \\ &= T_{0,i_k} \hat{s}_{i_k} T_{i_k,n[k]_k}, \end{aligned} \tag{4.33}$$

we have

$$dT_{0,n[k]_k} = \sum_{i=1}^{n[k]} T_{0,i_k} \hat{s}_{i_k} T_{i_k,n[k]_k} dq_{i_k}. \tag{4.34}$$

Left-multiplying both sides of Eq. (4.34) by $T_{0,n[k]_k}^{-1}$, then

$$T_{0,n[k]_k}^{-1} dT_{0,n[k]_k} = \sum_{i=1}^{n[k]} T_{i_k,n[k]_k}^{-1} \hat{s}_{i_k} T_{i_k,n[k]_k} dq_{i_k}. \tag{4.35}$$

Equation (4.35) is termed as the differential kinematic equation of branch k. Geometrically, it describes the differential change of the end link pose resulting from the differential changes of joint angles, viewing at the end link frame.

Let $T_{0,n[k]_k}^{d}$ denote the desired pose of the end link. When it is in the neighborhood of a nominal pose of the end link, denoted by $T_{0,n[k]_k}$, we have

$$dT_{0,n[k]_k} = T_{0,n[k]_k}^{d} - T_{0,n[k]_k}. \tag{4.36}$$

Therefore, the left side of Eq. (4.35) becomes

$$T_{0,n[k]_k}^{-1} dT_{0,n[k]_k} = T_{0,n[k]_k}^{-1} T_{0,n[k]_k}^{d} - I_{4\times 4}. \tag{4.37}$$

According to the definition of matrix logarithm [242],

$$\log(T_{0,n[k]_k}^{-1} T_{0,n[k]_k}^{d}) = T_{0,n[k]_k}^{-1} T_{0,n[k]_k}^{d} - I - \frac{(T_{0,n[k]_k}^{-1} T_{0,n[k]_k}^{d} - I)^2}{2} + \frac{(T_{0,n[k]_k}^{-1} T_{0,n[k]_k}^{d} - I)^3}{3} - \cdots . \tag{4.38}$$

Using first order approximation, we get

$$T_{0,n[k]_k}^{-1} dT_{0,n[k]_k} = \log(T_{0,n[k]_k}^{-1} T_{0,n[k]_k}^{d}). \tag{4.39}$$

Substituting Eq. (4.39) into Eq. (4.35), we obtain

$$\log(T_{0,n[k]_k}^{-1} T_{0,n[k]_k}^{d}) = \sum_{i=1}^{n[k]} T_{i_k,n[k]_k}^{-1} \hat{s}_{i_k} T_{i_k,n[k]_k} dq_{i_k}. \tag{4.40}$$

The explicit formulae for calculating the logarithm of elements of $SO(3)$ and $SE(3)$ are derived by Park [87] (refer to Sect. 4.2). They are the inverse of exponential map, i.e., log: $SO(3) \mapsto so(3)$ and $SE(3) \mapsto se(3)$. Here, $\log(T_{0,n[k]_k}^{-1} T_{0,n[k]_k}^{d})$ is an element of $se(3)$ so that it can be identified by a 6×1 vector denoted by $\log(T_{0,n[k]_k}^{-1} T_{0,n[k]_k}^{d})^{\vee}$ in which the first and later three elements represent the position and orientation differences between $T_{0,n[k]_k}$ and $T_{0,n[k]_k}^{d}$. The twist $\hat{s}_{i_k}$ can also be identified with a 6×1 vector s_{i_k}. Converting Eq. (4.40) into the adjoint representation [62, 87], we get

$$\log(T_{0,n[k]_k}^{-1} T_{0,n[k]_k}^{d})^{\vee} = \sum_{i=1}^{n[k]} Ad_{T_{i_k,n[k]_k}}^{-1} s_{i_k} dq_{i_k}. \tag{4.41}$$

Since $T_{0,n[k]_k} = T_{0,i_k} T_{i_k,n[k]_k}$, we get $Ad_{T_{i_k,n[k]_k}^{-1}} = Ad_{T_{0,n[k]_k}^{-1}} Ad_{T_{0,i_k}}$. Equation (4.41) can be rewritten as:

$$\log(T_{0,n[k]_k}^{-1} T_{0,n[k]_k}^{d})^{\vee} = Ad_{T_{0,n[k]_k}^{-1}} \sum_{i=1}^{n[k]} Ad_{T_{0,i_k}} s_{i_k} dq_{i_k}. \tag{4.42}$$

For simplicity, Eq. (4.42) can also be expressed in the following form:

$$D_{T_k} = J_k dq_k, \tag{4.43}$$

where
$D_{T_k} = \log(T_{0,n[k]_k}^{-1} T_{0,n[k]_k}^{d})^{\vee} \in \Re^{6\times 1}$, is referred to as the pose difference vector;
$J_k = A_k B_k S_k \in \Re^{6\times n[k]}$, is termed the body manipulator Jacobian matrix;
$A_k = Ad_{T_{0,n[k]_k}^{-1}} \in \Re^{6\times 6}$;
$B_k = row[Ad_{T_{0,1_k}}, Ad_{T_{0,2_k}}, \ldots, Ad_{T_{0,n[k]_k}}] \in \Re^{6\times 6n[k]}$;
$S_k = diag[s_{1_k}, s_{2_k}, \ldots, s_{n[k]_k}] \in \Re^{6n[k]_k \times n[k]}$;

$dq_k = column[dq_{1_k}, dq_{2_k}, \ldots, dq_{n[k]_k}] \in \Re^{n[k]_k \times 1}$.

Equation (4.43) is the differential inverse kinematic equation of a branch. We can get an inverse solution for a desired pose of the end link (an element of $SE(3)$) iteratively by using the Newton-Raphson method with Eq. (4.43). However, if the desired pose is not an element of $SE(3)$, rather, a pure orientation (an element of $SO(3)$) or a pure position (a 3×1 vector) problem, the inverse kinematic equation needs to be modified. For this purpose, we express D_{T_k} in a block matrix form as $D_{T_k} = [D_{P_k}, D_{R_k}]^T$, in which $D_{P_k} \in \Re^{3\times 1}$ and $D_{R_k} \in \Re^{3\times 1}$ represent the position difference vector and the orientation difference vector, respectively. Accordingly, $J_{T_k} = [J_{P_k}, J_{R_k}]^T$, in which $J_{P_k} \in \Re^{3\times n[k]}$ and $J_{R_k} \in \Re^{3\times n[k]}$ represent the body position Jacobian matrix and the body orientation Jacobian matrix, respectively. Equation (4.43) becomes

$$\begin{bmatrix} D_{P_k} \\ D_{R_k} \end{bmatrix} = \begin{bmatrix} J_{P_k} \\ J_{R_k} \end{bmatrix} dq_k. \tag{4.44}$$

1. **Pure orientation problem**
   A pure orientation problem means that the desired pose of the end link frame is an element of $SO(3)$, i.e., only the frame orientation is to be satisfied. Based on Eq. (4.44), we have

$$D_{R_k} = J_{R_k} dq_k, \tag{4.45}$$

   where
   $D_{R_k} = \log\left(R^{-1}_{0,n[k]_k} R^d_{0,n[k]_k}\right)^\vee = \log\left(R^T_{0,n[k]_k} R^d_{0,n[k]_k}\right)^\vee \in \Re^{3\times 1}$;
   $R^d_{0,n[k]_k} \in SO(3)$ is the desired orientation.
2. **Pure position problem**
   A pure position problem for a branch means that the desired pose of the end link frame is a 3×1 position vector, i.e., only the frame position (origin) is to be satisfied. Similar to Eq. (4.45), we have

$$D_{P_k} = J_{P_k} dq_k, \tag{4.46}$$

   where
   $D_{P_k} = R^{-1}_{0,n[k]_k}(P^d_{0,n[k]_k} - P_{0,n[k]_k}) = R^T_{0,n[k]_k}(P^d_{0,n[k]_k} - P_{0,n[k]_k}) \in \Re^{3\times 1}$;
   $P^d_{0,n[k]_k} \in \Re^{3\times 1}$ is the desired position.

4.4.2 Differential Kinematics Model for a Tree-Structured Robot

A tree-structured robot is a branching type mechanism without any loop. For such a robot, the inverse kinematics problem is to find a set of joint variable values that will place every end-effector to its own desired pose simultaneously. In general, several branches in a tree-structured robot may contain some common links and joints so that the inverse kinematic equations of these branches are dependent on each other. If we treat each of the branches as a serial type robot and individually calculate their inverse solutions, the computation results are usually infeasible because a common joint may be required to take different values at a time in order to place every end link to its own desired pose. Here, we propose an algorithm that can restrict any common joint to take only one value at a time automatically. This algorithm works by orderly combining the inverse kinematic equations of all constituting branches into a single equation according to the path matrix. When such an equation is solved numerically, a feasible solution can be derived. The input of this algorithm is a given AIM, a set of desired poses corresponding to all the end links, and an initial guess solution of the joint angles.

Based on the definition of the path matrix and Eq. (4.43), the combined inverse kinematic equation of a tree-structured robot can be given as follows:

$$D_T = J dq, \tag{4.47}$$

where
$D_T = column[D_{T_1}, D_{T_2}, \ldots, D_{T_m}] \in \Re^{6m\times 1}$, is termed as the generalized pose difference vector;
$J = ABS \in \Re^{6m\times n}$, is termed as the generalized body manipulator Jacobian matrix;
$A = diag[A_1, A_2, \ldots, A_m] \in \Re^{6m\times 6m}$;
$B = \begin{bmatrix} p_{12}Ad_{T_{0,1}} & p_{13}Ad_{T_{0,2}} & \cdots & p_{1(n+1)}Ad_{T_{0,n}} \\ p_{22}Ad_{T_{0,1}} & p_{23}Ad_{T_{0,2}} & \cdots & p_{2(n+1)}Ad_{T_{0,n}} \\ \cdots & & & \\ p_{m2}Ad_{T_{0,1}} & p_{m3}Ad_{T_{0,2}} & \cdots & p_{m(n+1)}Ad_{T_{0,n}} \end{bmatrix} \in \Re^{6m\times 6n}$, in which
p_{ij} $(i = 1, 2, \ldots, m; j = 1, 2, \ldots, n)$ is the entry in the ith row and jth column of path matrix $\mathcal{P}(\mathbf{G})$;
$S = diag[s_1, s_2, \ldots, s_n] \in \Re^{6n\times n}$;
$dq = column[dq_1, dq_2, \ldots, dq_n] \in \Re^{n\times 1}$.

For our purpose, the matrix J can also be written in a block matrix form of $J = column[J_1, J_2, \ldots, J_m]$ with $J_i = column[J_{P_i}, J_{R_i}]$ $(i = 1, 2, \ldots, m)$, where J_{P_i} and $J_{R_i} \in \Re^{3\times n}$.

Using the Newton-Raphson method, a closed-loop iterative algorithm similar to [17] is employed as shown in Fig. 4.4. This iterative algorithm determines the necessary changes in the joint angles to achieve the differential changes in the poses of the end links. Rewriting Eq. (4.47) in an iterative form, we get

$$dq^{i+1} = J^* D_T, \tag{4.48}$$

$$q^{i+1} = q^i + dq^{i+1}, \tag{4.49}$$

where i represents the number of iterations and J^* is the Moore-Pennose pseudoinverse of J. The Singular Values Decomposition (SVD) method is employed to

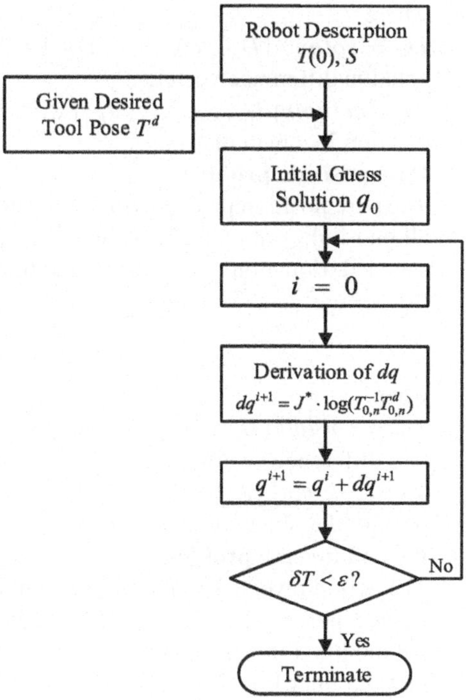

Fig. 4.4 Flow chart of iterative inverse kinematics algorithm

calculate J^*, which can treat non-square and singular Jacobian matrices in a uniform way [243].

Given an AIM and a set of desired poses T^d corresponding to each of the end links, the algorithm starts from an initial guess, q_0, somewhere in the neighborhood of the desired solution. It is ended when the norm of dq converges to a small tolerance ϵ and the desired poses for every branch thus are achieved simultaneously. Since the AIM contains all the necessary assembly information of a modular robot, the path matrix, forward kinematic equation, and inverse kinematic equation can be generated automatically once an AIM is given. Note that Eq. (4.47) works when the desired poses for each of the end link frames are all elements of $SE(3)$. However, the inverse kinematics problem of a general tree-structured robot may also be a pure orientation problem, a pure position problem, or a hybrid problem. In order to generalize this algorithm for various inverse kinematics problems, we make the following modifications:

1. **Pure orientation problem**
   The pure orientation problem for a tree-structured robot means that the desired poses of all end links are elements of $SO(3)$, i.e., only the end links' orientations are to be satisfied. Equation (4.47) can be modified as:

$$D_R = J_R dq, \tag{4.50}$$

where
$D_R = column[D_{R_1}, D_{R_2}, \ldots, D_{R_m}] \in \Re^{3m\times 1}$, is termed as the generalized orientation difference vector;
$J_R = column[J_{R_1}, J_{R_2}, \ldots, J_{R_m}] \in \Re^{3m\times n}$, is termed as the generalized body orientation Jacobian matrix.

2. **Pure position problem**
The pure position problem for a tree-structured robot means the desired poses of all end links are elements of $\Re^{3\times 1}$, i.e., only the end links' positions are to be satisfied. Equation (4.47) can be modified as:

$$D_P = J_P dq, \tag{4.51}$$

where
$D_P = column[D_{P_1}, D_{P_2}, \ldots, D_{P_m}] \in \Re^{3m\times 1}$, is termed as the generalized position difference vector;
$J_P = column[J_{P_1}, J_{P_2}, \ldots, J_{P_m}] \in \Re^{3m\times n}$, is termed as the generalized body position Jacobian matrix.

3. **Hybrid inverse problem**
The hybrid inverse kinematics problem for a tree-structured robot means that the desired poses set contains either elements of $SE(3)$, $SO(3)$, or $\Re^{3\times 1}$. Hence, the modified inverse kinematic equation should also be a mixed one containing part of either Eq. (4.47), (4.50), or (4.51) accordingly. As an example, consider a tree-structure modular robot with three branches. The desired poses for branch 1, branch 2, and branch 3 are elements of $SE(3)$, $SO(3)$, and $\Re^{3\times 1}$, respectively. The resulting inverse kinematic equation for such a hybrid problem can thus be given as follows:

$$D_H = J_H dq, \tag{4.52}$$

where
$D_H = column[D_{T_1}, D_{R_2}, D_{P_3}] \in \Re^{12\times 1}$,is termed as the hybrid pose difference vector, and $D_{T_1} \in \Re^{6\times 1}$, $D_{R_2}, D_{P_3} \in \Re^{3\times 1}$;
$J_H = column[J_1, J_{R_2}, J_{P_3}] \in \Re^{12\times n}$, is termed as the hybrid body manipulator Jacobian matrix, and $J_1 \in \Re^{6\times n}$, $J_{R_2}, J_{P_3} \in \Re^{3\times n}$, and n is the number of joints;
$dq = column[dq_1, dq_2, \ldots, dq_n] \in \Re^{n\times 1}$.

In order to illustrate the generality and effectiveness of the proposed inverse kinematics algorithm, two computation examples are presented in this section: one is the inverse kinematics of a tree-structured modular robot; the other is the inverse kinematics of a serial type robot. Note that in these examples, the joint angles are expressed in radians (for revolute joints) or millimeters (for prismatic joints), and the position vectors are in millimeters. The computation programs are coded in *Mathematica*, and executed on a PC with Pentium 133 processor.

4.4.3 Computation Examples

Example 4.3 As shown in Fig. 4.3, this modular robot has two branches and consists of 11 modules. According to the path matrix, branch 1 contains modules 0, 1, 2, 3, 4, 6, and 8 in which module 8 is the end link, while branch 2 contains modules 0, 1, 2, 3, 5, 7, 9, and 10 in which module 10 is the end link. Branch 1 is a 5-DOF (joint 4 is a fixed joint) RPRFRR type open chain, and branch 2 is a 7-DOF RPRRRRR type open chain. The common modules of these two branches are modules 0, 1, 2, and 3.

1. The desired poses of module 8 and module 10 are both elements of $SE(3)$. The desired poses are as follows:

$$T_{0,8}^d = \begin{bmatrix} -0.5 & -0.5 & -0.707107 & 358.471 \\ -0.5 & -0.5 & 0.707107 & 472.380 \\ -0.707107 & 0.707107 & 0 & 880.546 \\ 0 & 0 & 0 & 1 \end{bmatrix}, \text{ and}$$

$$T_{0,10}^d = \begin{bmatrix} 0.228553 & 0.780330 & 0.582107 & -22.097 \\ -0.875000 & 0.426777 & -0.228553 & 601.539 \\ -0.426777 & -0.457107 & 0.780330 & 1492.390 \\ 0 & 0 & 0 & 1 \end{bmatrix}.$$

Using our inverse kinematics algorithm, the numerical solution of q converges to $[0.7854, 100.00, 0.7854, 0, 0.7854, 0.7854, 0.7854, 0.7854, 0.7854, 0.7854\,]^T$ after 5 iterations with 5.38 s of computation time. The initial guess of q is $[0.5236, 75, 0.5236, 0, 0.5236, 0.5236, 0.5236, 0.5236, 0.5236, 0.5236\,]^T$. Putting q into the forward kinematics algorithm, the derived end link poses (actual poses) are the same as the desired ones. This indicates that the inverse kinematics solution is correct.

2. The desired pose of module 8 is a 3×1 vector, while the desired pose of module 10 is an element of $SO(3)$. The desired poses are as follows:

$$P_{0,8}^d = [358.471, 472.380, 880.546]^T, \text{ and}$$

$$R_{0,10}^d = \begin{bmatrix} 0.228553 & 0.780330 & 0.582107 \\ -0.875000 & 0.426777 & -0.228553 \\ -0.426777 & -0.457107 & 0.780330 \end{bmatrix}.$$

This is a hybrid inverse kinematics problem. The numerical solution of q converges to $[0.7854, 75.00, 0.7918, 0, 0.8745, 0.5236, 0.8914, 0.7175, 0.7046, 0.6917]^T$ after 5 iterations with 6.04 s of computation time. The initial guess of q is also $[0.5236, 75, 0.5236, 0, 0.5236, 0.5236, 0.5236, 0.5236, 0.5236, 0.5236]^T$. Using the forward kinematics algorithm, the actual poses corresponding to the numerical result of q are given by

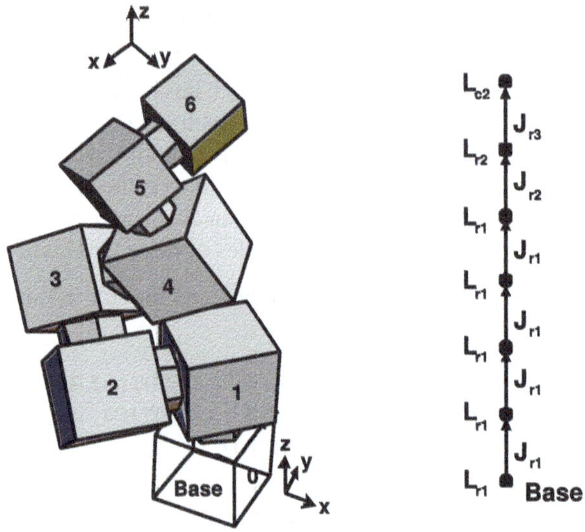

Fig. 4.5 A 6-DOF serial robot configuration

$$T^a_{0,8} = \begin{bmatrix} -0.412101 & -0.578610 & -0.703835 & 358.478 \\ -0.295000 & -0.646139 & 0.703903 & 472.378 \\ -0.862060 & 0.497711 & 0.095585 & 880.549 \\ 0 & 0 & 0 & 1 \end{bmatrix}, \text{ and}$$

$$T^a_{0,10} = \begin{bmatrix} 0.228529 & 0.780344 & 0.582097 & -61.974 \\ -0.875005 & 0.426756 & -0.228574 & 615.260 \\ -0.426779 & -0.457102 & 0.780331 & 1448.420 \\ 0 & 0 & 0 & 1 \end{bmatrix}.$$

The actual position of module 8, $P^a_{0,8}$, and the actual orientation of module 10, $R^a_{0,10}$, are also the same as the desired ones, respectively.

Example 4.4 As shown in Fig. 4.5, this modular robot consists of 7 modules, sequentially, module 0, 1, 2, 3, 4, 5, and 6 in which module 0 is the base and module 6 is the end link module. It is a 6-DOF RRRRRR type serial open chain. The given AIM and desired pose are

$$\mathcal{A}(G)=\begin{bmatrix} (z,y) & 0 & 0 & 0 & 0 & 0 & L_{r1} \\ (-x,y) & (z,y) & 0 & 0 & 0 & 0 & L_{r1} \\ 0 & (x,y) & (z,-x) & 0 & 0 & 0 & L_{r1} \\ 0 & 0 & (y,x) & (z,y) & 0 & 0 & L_{r1} \\ 0 & 0 & 0 & (x.-z) & (z,y) & 0 & L_{r1} \\ 0 & 0 & 0 & 0 & (-x,y) & (z,y) & L_{r2} \\ 0 & 0 & 0 & 0 & 0 & (x,y) & L_{c2} \\ J_{r1} & J_{r1} & J_{r1} & J_{r1} & J_{r2} & J_{r3} & 0 \end{bmatrix};$$

Table 4.1 Computation results

Cases	Contents	Algorithm based on POE formula	Algorithm based on D-H notation
1	Initial Guess	[0, 0, 0, 0, 0, 0]	[0, 0, 0, 0, 0, 0]
	Iterations	6	17
	Time (s)	2.20	5.98
	Solution	[−.4456, −.7854, 2.3562, −.4456, .7854, .7854]	[.5254, −.5826, 2.3747 − .8794, −.1670, .8787]
2	Initial Guess	[0.2, 0.2, 0.2, 0.2, 0.2, 0.2]	[0.2, 0.2, 0.2, 0.2, 0.2, 0.2]
	Iterations	14	96
	Time (s)	4.72	36.85
	Solution	[.5254, −.5826, 2.3747, −.8794, −.1670, .8787]	[.7854, −.4456, 2.3562, −.7854, −.4456, .7854]
3	Initial Guess	[0.4, 0.4, 0.4, 0.4, 0.4, 0.4]	[0.4, 0.4, 0.4, 0.4, 0.4, 0.4]
	Iterations	5	40
	Time (s)	1.87	17.96
	Solution	[.1827, .5799, .7672, .8783, 1.4156, .8779]	[−.4456, −.7854, 2.3562, −.4456, .7854, .7854]
4	Initial Guess	[0.6, 0.6, 0.6, 0.6, 0.6, 0.6]	[0.6, 0.6, 0.6, 0.6, 0.6, 0.6]
	Iterations	5	8
	Time (s)	1.76	2.86
	Solution	[.1827, .5799, .7672, .8783, 1.4156, .8779]	[.1827, .5799, .7672, .8783, 1.4156, .8779]
5	Initial Guess	[0.75, 0.75, 0.75, 0.75, 0.75, 0.75]	[0.75, 0.75, 0.75, 0.75, 0.75, 0.75]
	Iterations	3	4
	Time (s)	1.1	1.37
	Solution	[.7854, .7854, .7854, .7854, .7854, .7854]	[.7854, .7854, .7854, .7854, .7854, .7854]
6	Initial Guess	[0.8, 0.8, 0.8, 0.8, 0.8, 0.8]	[0.8, 0.8, 0.8, 0.8, 0.8, 0.8]
	Iterations	3	3
	Time (s)	0.99	0.93
	Solution	[.7854, .7854, .7854, .7854, .7854, .7854]	[.7854, .7854, .7854, .7854, .7854, .7854]

$$T^d_{0,6} = \begin{bmatrix} -0.426777 & 0.875000 & 0.228553 & -149.226 \\ -0.780330 & -0.228553 & -0.582107 & 115.235 \\ -0.457107 & -0.426777 & 0.780330 & 1164.930 \\ 0 & 0 & 0 & 1 \end{bmatrix}.$$

Actually, the desired pose is derived by putting a set of joint angles q =[0.7854, 0.7854, 0.7854, 0.7854, 0.7854, 0.7854]^T into the forward kinematics algorithm so that the inverse solutions must exist. In order to demonstrate the effectiveness and stability of our algorithm, a comparison with the method proposed by Kelma et al. [17] in computational aspects is given in Table 4.1. Note that Kelma's algorithm, based on the D-H notation, is also proposed for modular robots. Based on the Newton-Raphson method, this algorithm can also be used to derive the inverse kinematics solutions for serial type robots automatically. It is said to be optimized for computational efficiency and robustness, and has been implemented on a Motorola 68020/68881 based single board computer at rate on the order of 20 Hz.

4.4.4 Remarks on Computation Results

The first computation example is to illustrate the generality of the proposed algorithm. Hence, we choose a more general tree-structured modular robot. This robot, as shown in Fig. 4.3, consists of two branches with three common joints. Each branch contains both revolute joints and prismatic joints. Branch 1 is a 5-DOF nonredundant open chain, while branch 2 is a 7-DOF redundant one. What encouraged us is that the computation results are satisfactory in both cases, say, only 5 iterations are needed to obtain the desired solutions. This example also shows that the proposed algorithm is a general approach. It can deal with various inverse kinematics problems of a tree-structured modular robot automatically, regardless of joint types and DOFs.

The second example, on the other hand, is to demonstrate the computational efficiency and stability of this algorithm. As shown in Table 4.1, our algorithm always uses less iterations as well as computation durations than the method proposed in [17] does. As the initial guess solution is given close to the desired solution, the differences between these two algorithms become less distinct because both methods are locally convergent. This example shows that our algorithm is more efficient and stable even if the initial guess solution is not very close to any nominal solution. The possible reasons are: (1) instead of using the conventional Jacobian matrix, this inverse kinematic algorithm is formulated in terms of the body velocity Jacobian, and (2) without using the differential homogeneous transformation matrix [68], the differential change of the end link pose is computed by using the matrix logarithm on $SE(3)$. As a result, this newly developed algorithm might have a larger local convergent field. However, the exact reasons are still not clear. More research effort needs to be done in the future for an analytical explanation. Note that the algorithm in [17] is also implemented using *Mathematica*.

Chapter 5
Kinematic Calibration for Modular Serial Robots

Abstract This chapter addresses the kinematic calibration of a modular robot to compensate for the kinematic errors in the robot. Based on the local frame representation of the Product-of-Exponentials (POE) formula, a set of compact kinematic calibration models has been formulated for modular reconfigurable robots. The kinematic errors are attributed to the initial poses in the dyads so that the calibration models are significantly simplified. An iterative least square algorithm is employed for the calibration solution. Three corresponding simulation examples are demonstrated.

5.1 Introduction

In the modular robot system, a set of modules is joined together through connectors/adapters to form a complete robot assembly. However, due to the machining tolerance in the fabrication of the modules, the misalignments in the consecutive modules during assembly, and the wear and drift in the mechanical parts after a long period of use, geometric errors are introduced in the robot kinematic parameters. This will certainly lower the positioning accuracy of the robot. Compared to a monolithic designed industrial robot, a modular robot has more connecting points. Furthermore, its configuration will often be changed to cater for different task requirements through reselection and reconfiguration of the modules. Hence, identifying the critical kinematic parameters to improve the positioning accuracy of the end-effector is a very important issue for modular reconfigurable robots. Modular robot calibration comes into play in at least two situations. One is after the reconfiguration of a robot geometry. In this case, new kinematic parameters of the robot have to be identified and corrected. The other is after a large number of operations because of the possible loose module connections. The actual kinematic parameters may vary from those previously calibrated.

G. Yang and I.-M. Chen, *Modular Robots: Theory and Practice*, Research on Intelligent Manufacturing,
https://doi.org/10.1007/978-981-16-5007-9_5

Basically, kinematic calibration is necessary for all kinds of robots. In recent years, there has been an explosion in the literature on the topic of kinematic calibration, reflecting its current importance and development in robotics. Many calibration models have been presented by researchers who are motivated by the need for robot calibration [89–93]. Based on the different kinematic modeling techniques used, we categorized them into (1) the Denavit-Hartenberg (D-H) parameterization approach [93–99], (2) the zero reference position modeling approach [100–103], (3) the Continuous and Parametrically Complete (CPC) modeling approach [104–106], and (4) the Product-of-Exponentials (POE) formulation approach [23–26, 107–109].

The use of the D-H representation [56] for kinematics is the de-facto standard in robotics. This modeling technique intelligently uses a minimal set of parameters to describe the relationship between adjacent joint axes for the formulation of the kinematics. However, being adopted for calibration modeling, this method is not amenable to direct identification as all the parameters in this kinematic model are stringently defined and are unique to the particular robot configuration concerned [93]. In addition, it has the singularity problem when neighboring joint axes are nearly parallel. For example, one of the parameters in the D-H model is the length of the normal between two consecutive joint axes. If two axes are parallel, an infinite number of common normals exists with all having the same length. However, when one of the axes becomes slightly misaligned (with two axes still in a same plane), the length of this common normal will instantaneously jump to zero. This variation of kinematic parameters with respect to changes in joint axis being not continuous, may result in severe numerical difficulties during the parameter identification of the calibration process [92]. A number of researchers used modified forms of the D-H formulation to overcome the singularity problem. Hayati et al. [94–97] introduced an angular alignment parameter β_i in place of the d_i parameter in the D-H model to represent a small misalignment between two consecutive parallel axes, and an additional linear parameter for handling prismatic joints. Many researchers [110–113] pursued their research based on Hayati's modified D-H model. Stone [93] used a six-parameter representation (S-model) in which two additional parameters are added to the D-H model to allow for arbitrary placement of link frames.

Mooring [102], Mooring and Tang [103], and Kazerounian and Qian [100, 101] followed a so-called "zero-reference" position method as described in [58, 59]. This method described the robot kinematics in terms of the axes directions and locations in the zero reference position. Construction of a calibration model based on this method was not very straightforward. The elimination of redundant parameters in [103] required an assumption that the rotation and translation errors in certain directions are negligible. However, this technique is not prone to the difficulties encountered in case of parallel or nearly parallel joint axes when using the D-H notations.

Zhuang et al. [104–106] proposed a six-parameter CPC model. In this model, a singularity-free line representation consisting of four line parameters was adopted to ensure parameterically continuous, while two additional parameters were used to allow arbitrary placement of the link coordinate frames to make the model complete.

The POE formula is uniform in modeling robots with both revolute and prismatic joints, and provides a very structured parameterization for an open chain robot. Significantly, the kinematic parameters in the POE formula vary smoothly with changes in joint axes so that no special descriptions are required when adjacent joint axes are close to parallel. This fact makes the POE formula more suitable for calibration purpose. Park and Okamura [107, 108] proposed a calibration model based on the POE formula for general open chain robots. This model assumes that kinematic errors exist in each of the joint axes and the home position of the tool frame. All the kinematic parameters are expressed with respect to the base frame, and the attachment of local frames to each joint axis is unnecessary.

Because the robot geometries that a modular robot system can generate differ significantly, e.g., serial and tree-structured types, conventional calibration schemes dealing with specific type of robots with serial geometry are not suitable under such circumstances. Moreover, because of the modularity, it is simple and convenient to express the kinematic parameters in the local reference frames. Following the POE formula approach, we propose here a singularity-free kinematic calibration modeling method for the MRRS [23–26, 109]. This calibration model is based on the local frame representation of the POE formula, termed the Local POE model. In this modeling scheme, instead of identifying the kinematic errors in the joint axes relative to the module frames, which may involve complicated computations, all the kinematic errors are assumed to exist in the relative initial pose of every dyad. The calibration procedure is to redefine a new local coordinate frame on each module, in which the twist coordinates of the joint axis and joint angles retain their nominal values. By linearizing the forward kinematic equation with respect to the kinematic parameters in the local coordinate frames, four compact error models are derived for robot calibration. Then an iterative least-squares algorithm is employed for calibration solution. Three simulation examples of calibrating a 3-module robot, a SCARA type robot, and a tree-structured robot are demonstrated. The simulation results have shown that the average positioning accuracy of the end-effector increases significantly after calibration.

5.2 Kinematic Calibration Models

In general, the kinematic calibration model, i.e., the error model, can be derived by linearizing the forward kinematic equation with respect to the kinematic parameters. The kinematic calibration thus involves making several measurements of the end-effector pose error and determining the optimal (actual) values of the kinematic parameters to minimize the error in the least-square sense. For convenience, we first formulate the calibration model for serial type robots, and then modify the model to suit the calibration of tree-structured robots.

Consider a serial type robot with $n + 1$ modules and an end-effector (tool). Without loss of generality, we sequentially number the local coordinate frames from 0 (representing the base frame) to $n + 1$ (representing the tool frame). The forward kinematics can be given by

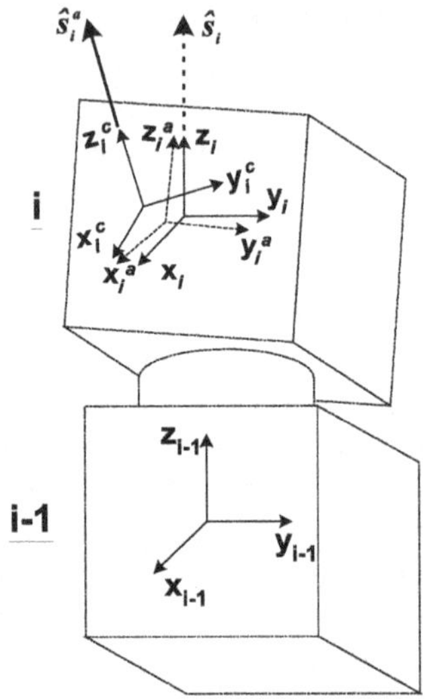

Fig. 5.1 Coordinate frames in a dyad

$$
\begin{aligned}
T_{0,n+1}(q) &= T_{0,1}(q_1)T_{1,2}(q_2)\ldots T_{n-1,n}(q_n)T_{n,n+1} \\
&= T_{0,1}(0)e^{\hat{s}_1 q_1}T_{1,2}(0)e^{\hat{s}_2 q_2}\ldots T_{n-1,n}(0)e^{\hat{s}_n q_n}T_{n,n+1},
\end{aligned} \tag{5.1}
$$

where $T_{n,n+1}$ represents the fixed kinematic transformation from the end module (end link) frame to the tool frame. The assembly of the end module and the tool can be considered as a fixed kinematic dyad, termed as the last dyad.

According to Eq. (5.1), the nominal forward kinematics $T_{0,n+1}(q)$ or simply $T_{0,n+1}$ is a function of $T(0)$, S, and q, denoted by $T = f(T(0), S, q)$, where $T(0)$, S, and q represent the initial poses, twist coordinates, and joint angles, respectively, i.e., $T(0) = [T_{0,1}(0), T_{1,2}(0),\ldots, T_{n-1,n}(0), T_{n,n+1}]^T$, $S = [s_1, s_2,\ldots,s_n]^T$, and $q = [q_1, q_2,\ldots,q_n]^T$.

We first consider two consecutive modules $(i-1)$ and i, i.e., a dyad, as shown in Fig. 5.1. Due to the geometric errors, module i's nominal frame i and twist $\hat{s}_i$ are shifted to the actual frame 'i^a' and twist '$\hat{s}_i^a$', respectively. The kinematic calibration, therefore, becomes to identify each of the actual module frames, joint twists, and the joint angle offsets. Of course, we can formulate an error model to directly identify these kinematic quantities. However, such a model will be very complicated because every dyad contains 13 redundant and highly coupled error parameters, i.e., six for the actual frame, six for the actual twist, and one for the actual joint angle. Especially, to identify the actual twists (joint axes) is rather difficult and may involve complicated computations [108].

By taking advantage of the local POE formula in which the local reference frames can be arbitrarily assigned, we are able to redefine a new local frame, termed the calibrated frame 'i^c' for each of modules during the calibration process. Then the robot positioning errors are minimized in the least-square sense, maintaining the twists and joint angles at their nominal values (Fig. 5.1) in the new defined local frames.

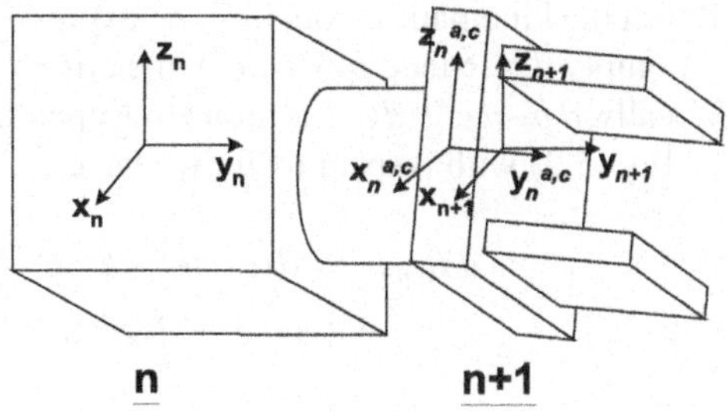

Fig. 5.2 Coordinate frames in the end-effector

As a result, the calibrated frame i^c may not coincide with the actual frame i^a. In this case, the new kinematic errors, termed the modeling errors, will be introduced to the dyad. Fortunately, the modeling errors in a dyad can be lumped into its succeeding dyad.

In the 'last dyad', i.e., the last module and the tool as shown in Fig. 5.2, there is no joint twist so that the geometric errors only exist in the initial pose. We, therefore, are able to recover the modeling errors presented in its preceding dyad and also achieve the coincidence of both the calibrated tool frame $(n+1)^c$ and the actual (measured) tool frame $(n+1)^a$ upon calibration. If we calibrate a robot without a tool, we can still add a tool frame that normally coincides with the last module frame. This additional dyad, termed as the compensation dyad, can be used to compensate for the modeling errors presented in its preceding dyad. Without loss of generality, we also assume that the base model does not have geometric errors, i.e., the nominal and actual base frames coincide with the world frame. Therefore, the kinematic calibration becomes to identify the new defined local frames so as to reflect the robot's actual characteristics.

5.2.1 *Basic Calibration Models*

According to the definition of matrix logarithm defined on $SE(3)$, we can find at least a $\hat{p} \in se(3)$ for a given $T \in SE(3)$, such that $e^{\hat{p}} = T$. Hence, it is sufficient to let $e^{\hat{p}_i} = T_{i-1,i}(0)$ (with $e^{\hat{p}_{n+1}} = T_{n,n+1}$), where $\hat{p}_i \in se(3)$ $(i = 1, 2, \ldots, n+1)$. Equation (5.1) can be rewritten as

$$T_{0,n+1} = e^{\hat{p}_1} e^{\hat{s}_1 q_1} e^{\hat{p}_2} e^{\hat{s}_2 q_2} \ldots e^{\hat{p}_n} e^{\hat{s}_n q_n} e^{\hat{p}_{n+1}}. \tag{5.2}$$

Assume that the kinematic errors occur only in $\hat{p}_i$, hence in $T_{i-1,i}(0)$ $(i = 1, 2, \ldots, n)$ and $T_{n,n+1}$. By linearizing the forward kinematic equation, Eq. (5.2) with respect to $\hat{p}_i$, the error models for robot calibration can be derived. Because the kinematic errors in $\hat{p}_i$ can be expressed in either frame i or frame $i-1$, the following two basic calibration models can be formulated accordingly.

1. Let the kinematic errors in $\hat{p}_i$ be expressed in module frame i, denoted by $d\hat{p}_{i,i}$ or simply $d\hat{p}_i$. Since $\hat{p}_i \in se(3)$, the error $d\hat{p}_i$ is also an element of $se(3)$. Geometrically, $d\hat{p}_i = e^{-\hat{p}_i} d(e^{\hat{p}_i})$ (refers to Appendix B) so that $d(e^{\hat{p}_i}) = e^{\hat{p}_i} d\hat{p}_i$. Linearizing Eq. (5.2) with respect to $\hat{p}_i$, we have

$$\begin{aligned}
dT_{0,n+1} &= d(e^{\hat{p}_1})e^{\hat{s}_1 q_1} e^{\hat{p}_2} e^{\hat{s}_2 q_2} \ldots e^{\hat{p}_n} e^{\hat{s}_n q_n} e^{\hat{p}_{n+1}} \\
&+ e^{\hat{p}_1} e^{\hat{s}_1 q_1} d(e^{\hat{p}_2}) e^{\hat{s}_2 q_2} \ldots e^{\hat{p}_n} e^{\hat{s}_n q_n} e^{\hat{p}_{n+1}} \\
&+ \ldots + e^{\hat{p}_1} e^{\hat{s}_1 q_1} \ldots e^{\hat{p}_{n-1}} e^{\hat{s}_{n-1} q_{n-1}} d(e^{\hat{p}_n}) e^{\hat{s}_n q_n} e^{\hat{p}_{n+1}} \\
&+ e^{\hat{p}_1} e^{\hat{s}_1 q_1} \ldots e^{\hat{p}_n} e^{\hat{s}_n q_n} d(e^{\hat{p}_{n+1}}) \\
&= e^{\hat{p}_1} d\hat{p}_1 e^{\hat{s}_1 q_1} e^{\hat{p}_2} e^{\hat{s}_2 q_2} \ldots e^{\hat{p}_n} e^{\hat{s}_n q_n} e^{\hat{p}_{n+1}} \\
&+ e^{\hat{p}_1} e^{\hat{s}_1 q_1} e^{\hat{p}_2} d\hat{p}_2 e^{\hat{s}_2 q_2} \ldots e^{\hat{p}_n} e^{\hat{s}_n q_n} e^{\hat{p}_{n+1}} \\
&+ \ldots + e^{\hat{p}_1} e^{\hat{s}_1 q_1} \ldots e^{\hat{p}_{n-1}} e^{\hat{s}_{n-1} q_{n-1}} e^{\hat{p}_n} d\hat{p}_n e^{\hat{s}_n q_n} e^{\hat{p}_{n+1}} \\
&+ e^{\hat{p}_1} e^{\hat{s}_1 q_1} \ldots e^{\hat{p}_n} e^{\hat{s}_n q_n} e^{\hat{p}_{n+1}} d\hat{p}_{n+1}. \qquad (5.3)
\end{aligned}$$

Right-multiplying both sides of the resulting equation by $T_{0,n+1}^{-1}$ and using the adjoint representation, the following error model can be shown:

$$\begin{aligned}
dT_{0,n+1}T_{0,n+1}^{-1} &= Ad_{e^{\hat{p}_1}} d\hat{p}_1 + Ad_{e^{\hat{p}_1} e^{\hat{s}_1 q_1}} Ad_{e^{\hat{p}_2}} d\hat{p}_2 \\
&+ \ldots + Ad_{e^{\hat{p}_1} e^{\hat{s}_1 q_1} e^{\hat{p}_2} e^{\hat{s}_2 q_2} \ldots e^{\hat{p}_n} e^{\hat{s}_n q_n}} Ad_{e^{\hat{p}_{n+1}}} d\hat{p}_{n+1}, \qquad (5.4)
\end{aligned}$$

where $dT_{0,n+1}T_{0,n+1}^{-1} \in se(3)$ represents the gross kinematic errors of a complete robot with respect to the base frame. According to the formulation of matrix logarithm as mentioned in Chap. 4, $dT_{0,n+1}T_{0,n+1}^{-1}$ can be given by

$$dT_{0,n+1}T_{0,n+1}^{-1} = \log(T_{0,n+1}^{a} T_{0,n+1}^{-1}), \qquad (5.5)$$

where $T_{0,n+1}^{a}$ represents the actual (measured) pose of the tool frame with respect to the base frame. Since elements of $se(3)$ can also be identified by 6-dimensional vectors through: $d\hat{p}_i \mapsto dp_i = (dx_i, dy_i, dz_i, \delta x_i, \delta y_i, \delta z_i)^T \in \Re^{6\times 1}$ and $\log(T_{0,n+1}^{a} T_{0,n+1}^{-1}) \mapsto \log(T_{0,n+1}^{a} T_{0,n+1}^{-1})^{\vee} = (dx_{0,n+1}, dy_{0,n+1}, dz_{0,n+1}, \delta x_{0,n+1}, \delta y_{0,n+1}, \delta z_{0,n+1})^T \in \Re^{6\times 1}$, Eq. (5.4) can be expressed as:

$$\begin{aligned}
\log(T_{0,n+1}^{a} T_{0,n+1}^{-1})^{\vee} &= Ad_{T_{0,1}(0)} dp_1 + Ad_{T_{0,1}} Ad_{T_{1,2}(0)} dp_2 \\
&+ \ldots + Ad_{T_{0,n}} Ad_{T_{n,n+1}(0)} dp_{n+1}, \qquad (5.6)
\end{aligned}$$

where $T_{0,i} = e^{\hat{p}_1} e^{\hat{s}_1 q_1} e^{\hat{p}_2} e^{\hat{s}_2 q_2} \ldots e^{\hat{p}_i} e^{\hat{s}_i q_i}$, representing the forward kinematic transformation from the base frame 0 to module frame i. After dp_i is identified, the calibrated initial pose of module frame i with respect to module frame $i-1$, $T_{i-1,i}^{c}(0)$, can be given by

$$T^c_{i-1,i}(0) = e^{\hat{p}_i} e^{d\hat{p}_i} = T_{i-1,i}(0) e^{d\hat{p}_i}. \quad (5.7)$$

2. Let the kinematic errors in $\hat{p}_i$ be expressed in the module frame $i-1$, denoted by $d\hat{p}_{i-1,i}$. Then $d\hat{p}_{i-1,i} \in se(3)$ and $d\hat{p}_{i-1,i} = d(e^{\hat{p}_i})e^{-\hat{p}_i} = Ad_{T_{i-1,i}(0)} d\hat{p}_i$. Equation (5.6) can be rewritten as

$$\log(T^a_{0,n+1} T^{-1}_{0,n+1})^\vee = dp_{0,1} + Ad_{T_{0,1}} dp_{1,2} + \ldots + Ad_{T_{0,n}} dp_{n,n+1}, \quad (5.8)$$

where $dp_{i-1,i} = (dx_{i-1,i}, dy_{i-1,i}, dz_{i-1,i}, \delta x_{i-1,i}, \delta y_{i-1,i}, \delta z_{i-1,i})^T \in \Re^{6\times 1}$. After $dp_{i-1,i}$ is identified, the calibrated initial pose of module frame i with respect to module frame $i-1$, $T^c_{i-1,i}(0)$, can be given by

$$T^c_{i-1,i}(0) = e^{d\hat{p}_i} e^{\hat{p}_i} = e^{d\hat{p}_i} T_{i-1,i}(0). \quad (5.9)$$

The basic error models of Eqs. (5.6) and (5.8) are derived by observing the gross kinematic errors of a complete robot with respect to the base frame. Alternatively, the gross kinematic errors of the robot can also be expressed with respect to the tool frame. Left-multiplying both sides of Eqs. (5.6) and (5.8) by $Ad_{T^{-1}_{0,n+1}}$, the following two equations can be easily shown:

$$\begin{aligned}\log(T^{-1}_{0,n+1} T^a_{0,n+1})^\vee = Ad_{T^{-1}_{0,n+1}} (&Ad_{T_{0,1}(0)} dp_1 + Ad_{T_{0,1}} Ad_{T_{1,2}(0)} dp_2 \\ &+ \ldots + Ad_{T_{0,n}} Ad_{T_{n,n+1}(0)} dp_{n+1});\end{aligned} \quad (5.10)$$

$$\log(T^{-1}_{0,n+1} T^a_{0,n+1})^\vee = Ad_{T^{-1}_{0,n+1}} (dp_{0,1} + Ad_{T_{0,1}} dp_{1,2} + \ldots + Ad_{T_{0,n}} dp_{n,n+1}); \quad (5.11)$$

where $\log(T^{-1}_{0,n+1} T^a_{0,n+1})^\vee$ is a 6-dimensional vector associated with $\log(T^{-1}_{0,n+1} T^a_{0,n+1}) \in se(3)$, representing the gross kinematic errors of the robot with respect to the tool frame, and $Ad_{T^{-1}_{0,n+1}} \log(T^a_{0,n+1} T^{-1}_{0,n+1}) = \log(T^{-1}_{0,n+1} T^a_{0,n+1}) = T^{-1}_{0,n+1} dT_{0,n+1}$.

By expressing the local and gross kinematic errors of a robot with respect to different reference frames, four compact calibration models are formulated, i.e., Eqs. (5.6), (5.8), (5.10), and (5.11). For the purpose of kinematic calibration, these models are equivalent. From the computational point of view, however, these four models are slightly different. For example, the term $\log(T^{-1}_{0,n+1} T^a_{0,n+1})^\vee$ in both Eqs. (5.10) and (5.11) can be calculated in a simple and efficient way such that:

$$\log(T^{-1}_{0,n+1} T^a_{0,n+1})^\vee = \begin{bmatrix} (R - R^T)/2 & R^T_{0,n+1}(P^a_{0,n+1} - P_{0,n+1}) \\ 0 & 0 \end{bmatrix}, \quad (5.12)$$

where $T^a_{0,n+1}$ and $T_{0,n+1}$ are given by $(P^a_{0,n+1}, R^a_{0,n+1})$ and $(P_{0,n+1}, R_{0,n+1})$, respectively. In Eq. (5.12), $R = R^T_{0,n+1} R^a_{0,n+1}$. Geometrical interpretations of the kinematic terms such as $d\hat{p}_i$, $d\hat{p}_{i-1,i}$, $d(e^{\hat{p}_i})e^{-\hat{p}_i}$, $e^{-\hat{p}_i} d(e^{\hat{p}_i})$, $dT_{0,n+1} T^{-1}_{0,n+1}$, and $T^{-1}_{0,n+1} dT_{0,n+1}$, can be found in Appendix B.

5.2.2 An Iterative Least-Squares Algorithm

Based on the calibration models, an iterative least-squares algorithm is employed for the calibration solution. This algorithm is suitable for each of the calibration models. For simplicity, we will introduce this algorithm according to the error model of Eq. (5.6).

Apparently, Eq. (5.6) can also be expressed as a linear equation of the form

$$y = Ax, \tag{5.13}$$

where

$$y = \log(T^a_{0,n+1} T^{-1}_{0,n+1})^\vee \in \Re^{6\times 1};$$
$$x = \left[dp_1, dp_2, \ldots, dp_{n+1}\right]^T \in \Re^{6(n+1)\times 1};$$
$$A = \left[Ad_{T_{0,1}(0)}, Ad_{T_{0,1}} Ad_{T_{1,2}(0)}, \ldots, Ad_{T_{0,n}} Ad_{T_{n,n+1}(0)}\right] \in \Re^{6\times 6(n+1)}.$$

$T^{-1}_{0,n+1}$ and A can be obtained from the nominal model, while $T^a_{0,n+1}$ can be obtained from the actual measurement data. Hence, $y = \log(T^a_{0,n+1} T^{-1}_{0,n+1})^\vee$ can be computed. x is the kinematic errors to be identified, which reflects the errors in the robot.

The model presented above identifies six kinematic error parameters to calibrate the initial pose of each dyad. A n-DOF modular robot will totally have $6(n+1)$ error parameters to be identified. This number is greater than that of the conventional calibration model in which the number of error parameters is equal to $4R + 2P$, where R and P represent the number of revolute and prismatic joints, respectively [91]. Consequently, the A matrix in Eq. (5.13) is larger than that of the conventional model, which will make the computation heavy. Since the calibration computation is usually an off-line computation scheme and the A matrix, in general, is not very large, the computation complexity of the calibration models becomes less important.

The identification of the kinematic errors requires comparing the difference between the robot's actual (measured) and nominal poses. To improve the calibration accuracy, we usually need to measure the tool frame in many different robot postures. Suppose we need to take m measured pose data. For the ith measurement, we obtain a y_i. The corresponding A_i can be determined from the nominal kinematic model. After m measurements, we can stack y_i and A_i to form the following equation:

$$\widetilde{Y} = \widetilde{A}x, \tag{5.14}$$

where

$$\widetilde{Y} = [y_1, y_2, \ldots, y_m]^T \in \Re^{6m\times 1};$$
$$x = \left[dp_1, dp_2, \ldots, dp_{n+1}\right]^T \in \Re^{6(n+1)\times 1};$$
$$\widetilde{A} = column[A_1, A_2, \ldots, A_m] \in \Re^{6m\times 6(n+1)}.$$

Obviously, Eq. (5.14) is a linear calibration model. Since the model consists of $6m$ linear equations with $6(n+1)$ variables ($m > (n+1)$), the linear least-squares algorithm is employed for the parameter identification. The least-square solution of x is given by

$$x = (\widetilde{A}^T \widetilde{A})^{-1} \widetilde{A}^T \widetilde{Y}, \tag{5.15}$$

where $(\widetilde{A}^T \widetilde{A})^{-1} \widetilde{A}^T$ is the pseudoinverse of $\widetilde{A}$.

The solution of Eq. (5.15) can be further improved through iterative substitution as shown in Fig. 5.3. Let $T(0)$ denote the vector of the initial poses of the dyads and initialize $T(0)$ using the nominal value, i.e., $T(0) = [T_{0,1}(0), T_{1,2}(0), \ldots, T_{n-1,n}(0), T_{n,n+1}]^T$. Based on the measured data and Eq. (5.15), we can obtain the kinematic error parameter vector, x. The vector $T(0)$ is updated by substituting x into Eq. (5.7). The same procedure is repeated until the norm of the error vector, $||x||$, approaches zero and the vector $T(0)$ converges to some stable values. Then the final $T(0)$ represents the calibrated initial poses of the dyads, i.e., $T(0) =[T^c_{0,1}(0), T^c_{1,2}(0), \ldots, T^c_{n,n+1}(0)]^T$.

Note that the kinematic error vector, x, will no longer represent the actual kinematic errors in the dyads after iterations. However, the actual kinematic errors in each of the dyads can be extracted by comparing $T_{i-1,i}(0)$ with $T^c_{i-1,i}(0)$. Based on the matrix logarithm, Eq. (5.7) can also be written as

$$dp_i = \log(T^{-1}_{i-1,i}(0) T^c_{i-1,i}(0))^{\vee}. \tag{5.16}$$

After calibration, the forward kinematic equation becomes

$$T_{0,n+1} = T^c_{0,1}(0) e^{\hat{s}_1 q_1} T^c_{1,2}(0) e^{\hat{s}_2 q_2} \ldots T^c_{n-1,n}(0) e^{\hat{s}_n q_n} T^c_{n,n+1}. \tag{5.17}$$

5.2.3 *Kinematic Calibration of Tree-structured Robots*

All the calibration models mentioned above are based on the local POE formula. They can be easily modified to suit the calibration of a tree-structured modular robot. Since the branches in a tree-structured modular robot may contain several common modules (or dyads), all these branches are to be calibrated simultaneously in order to get a unique calibrated initial pose for each of the common dyads. Similar to the modeling scheme for inverse kinematics, once the path matrix is derived from a given AIM, the calibration model can be automatically constructed.

Now we consider a tree-structured modular robot consisting of $n + 1$ modules and m branches. The path matrix of such a robot therefore has m rows and $n + 1$ columns. Identical to the labeling scheme adopted in Sect. 4.3, we assume that row k ($k = 1, 2, \ldots, m$) has $n[k] + 1$ non-zero entries, sequentially 0_k or simply 0 (because of the unique base), $1_k, \ldots, n[k]_k$ from the left to right. Here i_k is the index of the ith non-zero entry in row k, and $i = 0, 1, \ldots, n[k]$. Since the tools that will be rigidly

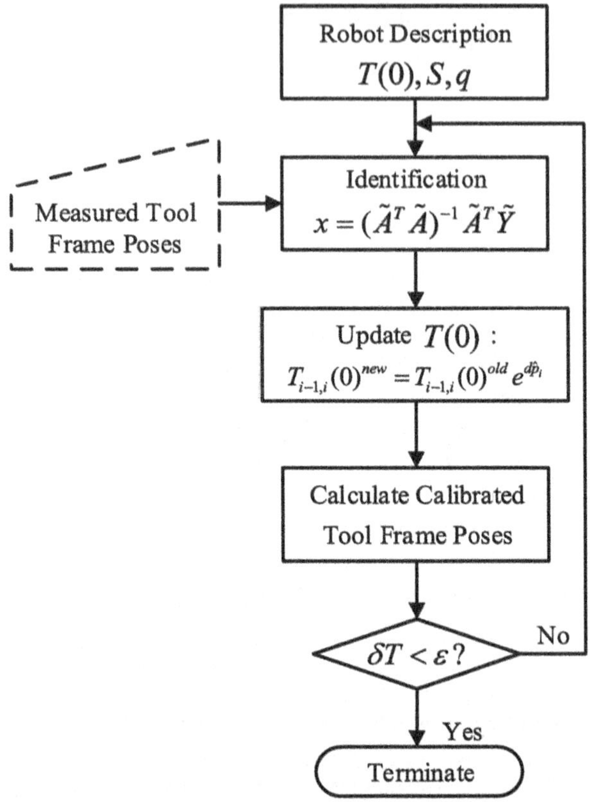

Fig. 5.3 Iterative calibration algorithm

attached to the pendant links are not considered in the AIM and the path matrix, the last dyad or the compensation dyad will be added to each branch. Based on Eq. (5.14), the calibration equation for branch k can be given by:

$$Y_k = A_k x_k, \tag{5.18}$$

where

$Y_k = \log(T^a_{0,n[k]+1} T^{-1}_{0,n[k]+1})^\vee \in \Re^{6\times 1}$;

$x_k = \left[dp_{1_k}, dp_{2_k}, \ldots, dp_{n[k]_k}, dp_{(n[k]+1)_k}\right]^T \in \Re^{6(n[k]+1)\times 1}$;

$A_k = \left[Ad_{T_{0,1_k}(0)}, Ad_{T_{0,1_k}} Ad_{T_{1_k,2_k}(0)}, \ldots, Ad_{T_{0,(n[k]-1)_k}} Ad_{T_{(n[k]-1)_k,n[k]_k}(0)}, Ad_{T_{0,n[k]_k}}\right]$

$\in \Re^{6\times 6(n[k]+1)}$.

The formulation of the calibration model for a tree-structured modular robot is similar to that of the inverse kinematics model, i.e., arranging the calibration equations of all branches into a single equation according to the path matrix. Instead of presenting the formulation details, we use an example to show the arranging scheme. As shown in Fig. 5.4, the tree-structured modular robot consists of 9 modules (7-DOFs). The AIM and path matrix are given as follows:

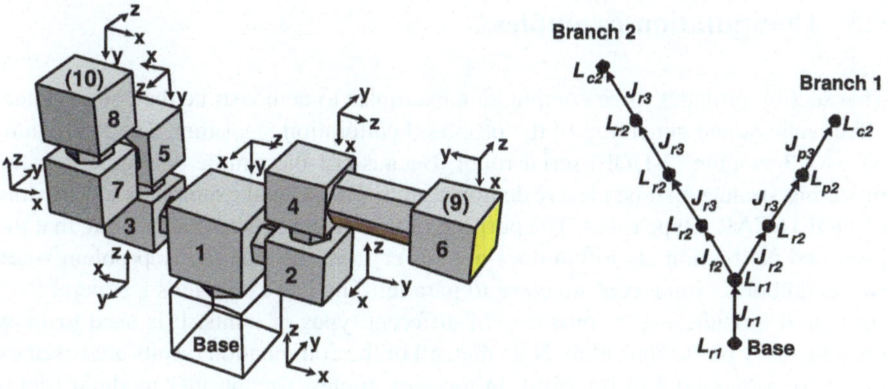

Fig. 5.4 A tree-structured modular robot configuration (7-DOF)

$$
\mathcal{A}(\tilde{G}) = \begin{bmatrix}
(z,-y) & 0 & 0 & 0 & 0 & 0 & 0 & 0 & L_{r1} \\
(y,x) & (z,x) & (-z,x) & 0 & 0 & 0 & 0 & 0 & L_{r1} \\
0 & (-y,x) & 0 & (z,x) & 0 & 0 & 0 & 0 & L_{r2} \\
0 & 0 & (-x,y) & 0 & (z,y) & 0 & 0 & 0 & L_{r2} \\
0 & 0 & 0 & (x,-y) & 0 & (z,x) & 0 & 0 & L_{p2} \\
0 & 0 & 0 & 0 & (y,z) & 0 & (z,x) & 0 & L_{r2} \\
0 & 0 & 0 & 0 & 0 & (y,x) & 0 & 0 & L_{c2} \\
0 & 0 & 0 & 0 & 0 & 0 & (y,-x) & (z,x) & L_{r2} \\
0 & 0 & 0 & 0 & 0 & 0 & 0 & (y,x) & L_{c2} \\
J_{r1} & J_{r2} & J_{f2} & J_{r3} & J_{r3} & J_{p3} & J_{r3} & J_{r3} & 0
\end{bmatrix};
$$

$$
\mathcal{P}(\mathbf{G}) = \begin{bmatrix} 1\,1\,1\,0\,1\,0\,1\,0\,0 \\ 1\,1\,0\,1\,0\,1\,0\,1\,1 \end{bmatrix}.
$$

According to the path matrix, we know that this robot has two branches. The first branch consists of link modules 0, 1, 2, 4 and 6; the second branch consists of link modules 0, 1, 3, 5, 7 and 8. Since the tools or end-effectors are not attached to each of the pendant links, a compensation dyad will be added to each branch. Based on the path matrix $\mathcal{P}(\mathbf{G})$ and Eq. (5.18), the calibration model for this robot can be written as:

$$
Y = Ax, \tag{5.19}
$$

where

$Y = \begin{bmatrix} Y_1 \\ Y_2 \end{bmatrix} = \begin{bmatrix} \log(T^a_{0,9} T^{-1}_{0,9})^\vee \\ \log(T^a_{0,10} T^{-1}_{0,10})^\vee \end{bmatrix} \in \Re^{12\times 1};$

$x = [dp_1, dp_2, \ldots, dp_8, dp_9, dp_{10}]^T \in \Re^{60\times 1};$

$$
A = \begin{bmatrix} B_{0,1} & A_{0,1}B_{1,2} & 0 & A_{0,2}B_{2,4} & 0 & A_{0,4}B_{4,6} & 0 & 0 & A_{0,6} & 0 \\ B_{0,1} & 0 & A_{0,1}B_{1,3} & 0 & A_{0,3}B_{3,5} & 0 & A_{0,5}B_{5,7} & A_{0,7}B_{7,8} & 0 & A_{0,8} \end{bmatrix}
$$

$\in \Re^{12\times 60}$, in which $A_{i,j} = Ad_{T_{i,j}}$ and $B_{i,j} = Ad_{T_{i,j}(0)}$.

This example shows that the proposed calibration models are also capable of calibrating robots with tree-structured geometry that most of the multi-arm robots belong to. Therefore, the local POE calibration model is a completely general approach.

5.3 Computation Examples

This section provides three computation examples to demonstrate the convergence, effectiveness, and generality of the proposed calibration algorithm. The first example is a 3-module (2-DOF) serial robot. Because of the simple configuration, the modeling strategy can be clearly demonstrated. The second example is a 7-module (4-DOF) SCARA type robot. The purpose of this example is to demonstrate that the proposed calibration algorithm does not suffer from the singularity problem when two neighboring joint axes are close to parallel. The last example is a general tree-structured modular robot consisting of different types of joints. It is used to show the generality of the algorithm. Note that, all of the computation results are based on the calibration model of Eq. (5.6). In these examples, we consider modular robots without any end-effector or tool.

5.3.1 *Calibration of a three-module Robot*

As shown in Fig. 5.5, the given modular robot consists of two large revolute joint modules and one small cubic link module. Its AIM is given as follows:

$$\mathcal{A}(\tilde{G}) = \begin{bmatrix} (z,x) & 0 & L_{r1} \\ (-x,y) & (z,y) & L_{r1} \\ 0 & (-z,x) & L_{c2} \\ J_{r1} & J_{r2} & 0 \end{bmatrix}.$$

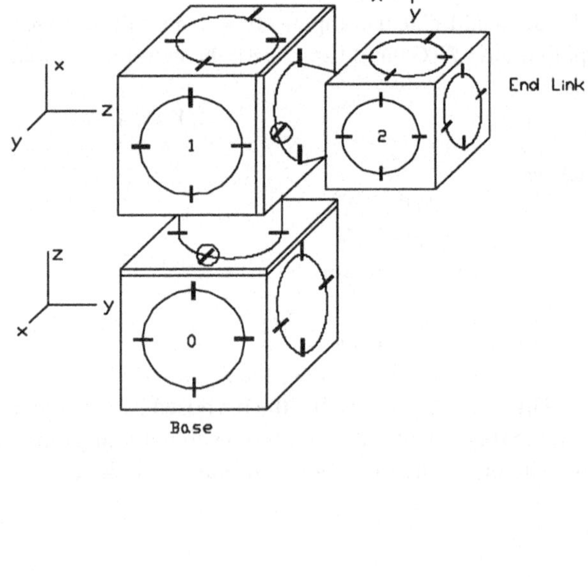

Fig. 5.5 A 3-module robot configuration (2-DOF)

According to the AIM, we find that:
$s_1 = (0, 0, 0, 1, 0, 0)^T$; $s_2 = (0, 0, 0, 0, 0, 1)^T$;

$$T_{0,1}(0) = \begin{bmatrix} 0 & 1 & 0 & 0 \\ 0 & 0 & 1 & 0 \\ 1 & 0 & 0 & 350 \\ 0 & 0 & 0 & 1 \end{bmatrix}; T_{1,2}(0) = \begin{bmatrix} 0 & -1 & 0 & 0 \\ 1 & 0 & 0 & 0 \\ 0 & 0 & 1 & 312.5 \\ 0 & 0 & 0 & 1 \end{bmatrix}.$$

Note that all the angles are expressed in radians, and the lengths are in millimeters. To validate the proposed calibration models, we present here a simulated calibration result. The measured (simulated) end link frame poses are obtained from Eq. (5.1) (with $T_{n,n+1} = I$) under a set of preset geometric errors. This procedure is stated as follows.

- Assign the geometric errors:

  dp_i, dq_i, and ds_i (i = 1, 2) (listed in Table 5.1);
- Find the actual initial pose in each dyad:

  $T^a_{i-1,i}(0) = T_{i-1,i}(0)e^{d\hat{p}_i}$ (i = 1, 2);
- Calculate the simulated end link frame poses:

  $T^a_{0,2}(q_1 + dq_1, q_2 + dq_2) = T^a_{0,1}(0)e^{(\hat{s}_1 + d\hat{s}_1)(q_1 + dq_1)}T^a_{1,2}(0)e^{(\hat{s}_2 + d\hat{s}_2)(q_2 + dq_2)}$.

In this example, the number of measured poses is set to 10, i.e, $m = 10$. The calibration results are given in Tables 5.2, 5.3 and Fig. 5.6. Table 5.2 shows the identified kinematic errors in the dyads. Because there is no end-effector or tool in this robot, the compensation term is also identified and given in the last row of Table 5.2. It is apparent that the preset geometric errors and the kinematic errors to be identified have different physical meanings and are not one-to-one correspondent so that the preset geometric errors can not be fully recovered. Nevertheless, as shown in Table 5.3, two end link poses (arbitrarily selected) after calibration are the same as the measured ones. It follows that these two sets of error parameters are equivalent in the sense of describing the actual end link poses. In robot calibration, the positioning accuracy is the main concern. Whether the error parameters can be fully recovered becomes less important. Figure 5.6 also shows that the calibrated (updated nominal) end link poses $T_{0,n} = (p_{0,n}, R_{0,n})$ fully converge to the measured ones $T^a_{0,n} = (p^a_{0,n}, R^a_{0,n})$ after only a few iterations. Note that the position and orientation deviations are given by $dp = \sqrt{\frac{\sum_{i=1}^m \|p^a_{0,n} - p_{0,n}\|}{m}}$ and $dR = \sqrt{\frac{\sum_{i=1}^m \|\log(R^{a\,-1}_{0,n} R_{0,n})^\vee\|}{m}}$, where dp and dR can be interpreted as the average position and orientation differences between the measured (actual) end link poses and the calibrated end link poses for m measurements, respectively. The computation results indicate that the proposed algorithm is correct and effective.

Now we consider whether the actual twists expressed in the calibrated module frames retain their nominal coordinates as we expected. The coordinates of the actual twists expressed in the local actual module frames can be given by:

Table 5.1 Preset geometric errors for the three-module robot

	dp_i	ds_i	dq_i
Dyad1	$(2., 2., 2., 0.02, 0.02, 0.02)^T$	$(0, 0, 2., -0.0002, 0.02, 0)^T$	0.02
Dyad2	$(2., 2., 2., 0.02, 0.02, 0.02)^T$	$(2., 0, 0, 0, 0.02, -0.0002)^T$	0.02

Table 5.2 Identified kinematic errors for the three-module robot

	dp_i
Dyad1	$(-0.961, -0.073, 2.010, 0.0272, 0.0201, 0.0397)^T$
Dyad 2	$(0.009, 0.939, 1.013, 0.0001, 0.0072, 0.0270)^T$
Compensation Dyad	$(-0.007, -1.990, 1.033, 0.0200, -0.0001, -0.0070)^T$

$s_1^a = s_1 + ds_1 = (0, 0, 2., .9998, .02, 0)^T$ and
$s_2^a = s_2 + ds_2 = (2., 0, 0, 0, .02, .9998)^T$.

According to the preset errors, the actual module frames with respect to the base frame can be computed as

$$T_{0,1}^a(0) = \begin{bmatrix} .0200 & .9990 & -.0396 & 2.000 \\ -.0204 & .0400 & .9990 & 2.040 \\ .9996 & -.0192 & .0212 & 352.001 \\ 0 & 0 & 0 & 1 \end{bmatrix};$$

$$T_{0,2}^a(0) = \begin{bmatrix} .9980 & -.0603 & -.0184 & -8.448 \\ .0208 & .0392 & .9990 & 316.345 \\ -.0596 & -.9974 & .0403 & 356.662 \\ 0 & 0 & 0 & 1 \end{bmatrix}.$$

From the calibration results, the calibrated module frames with respect to the base frame can also be computed as

$$T_{0,1}^c(0) = \begin{bmatrix} .0400 & .9988 & -.0268 & -.119 \\ -.0196 & .0276 & .9994 & 2.018 \\ .9990 & -.0394 & .0207 & 349.061 \\ 0 & 0 & 0 & 1 \end{bmatrix};$$

$$T_{0,2}^c(0) = \begin{bmatrix} .9976 & -.0669 & -.0196 & -8.548 \\ .0209 & .0190 & .9996 & 315.370 \\ -.0665 & -.9976 & .0204 & 354.605 \\ 0 & 0 & 0 & 1 \end{bmatrix}.$$

Therefore, the actual twists expressed in the calibrated module frames, denoted by $s_i^a[c]$ (i=1, 2), can be given by:

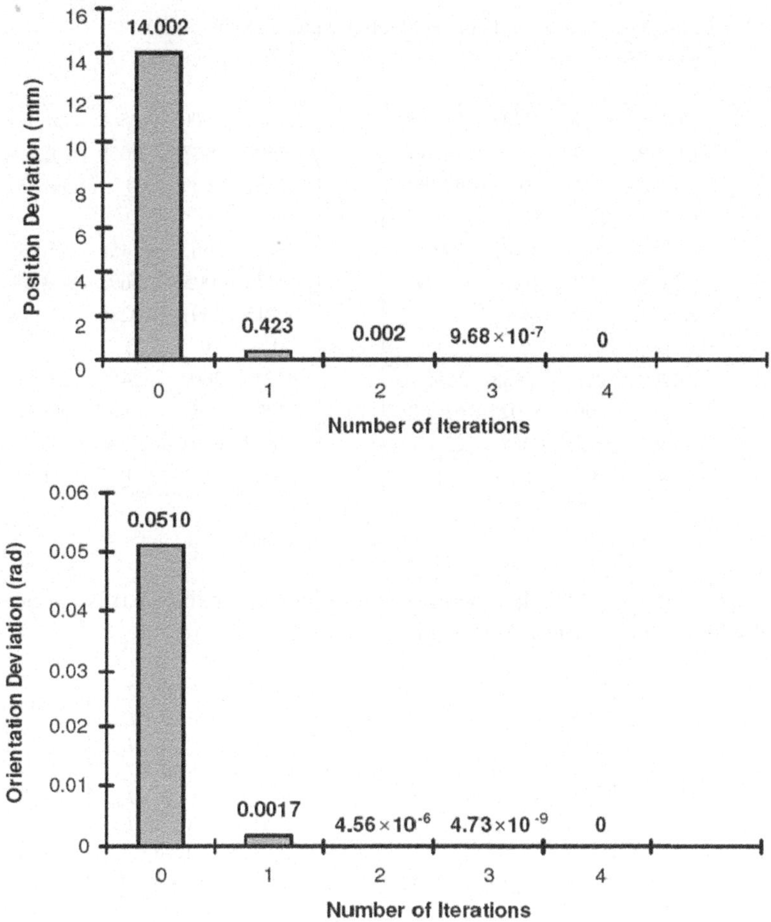

Fig. 5.6 Calibration convergence (three-module robot)

$s_1^a[c]= Ad^{-1}_{T^c_{0,1}(0)} Ad_{T^a_{0,1}(0)} s_1^a=(0, 0, 0, 1, 0, 0)^T=s_1$ and
$s_2^a[c]= Ad^{-1}_{T^c_{0,2}(0)} Ad_{T^a_{0,2}(0)} s_2^a=(0, 0, 0, 0, 0, 1)^T=s_2$.

The results also show that the actual twists expressed in the calibrated module frames retain their nominal coordinates so that our initial assumptions are valid.

5.3.2 Calibration of a SCARA Type Robot

A 4-DOF (RPFRFR) SCARA type modular robot is shown in Fig. 5.7. This robot consists of seven serially connected modules of different types, i.e., L_{r1}, L_{p1}, L_{c1},

Table 5.3 End link poses before and after calibration (three-module robot)

Joint Angle	$q = (.9948, -1.5184)^T$	$q = (2.8798, 2.8798)^T$
Nominal Pose	$\begin{bmatrix} .0285 & .5439 & -.8387 & -262.085 \\ .0439 & .8375 & .5446 & 170.200 \\ .9986 & -.0523 & 0 & 350.000 \\ 0 & 0 & 0 & 1 \end{bmatrix}$	$\begin{bmatrix} .9330 & .2500 & -.2588 & -80.881 \\ -.2500 & -.0670 & -.9659 & -301.852 \\ -.2588 & .9659 & 0 & 350.000 \\ 0 & 0 & 0 & 1 \end{bmatrix}$
Measured Pose	$\begin{bmatrix} .0780 & .5063 & -.8588 & -269.263 \\ .0419 & .8590 & .5102 & 164.279 \\ .9961 & -.0758 & .0458 & 362.220 \\ 0 & 0 & 0 & 1 \end{bmatrix}$	$\begin{bmatrix} .9322 & .2662 & -.2450 & -73.885 \\ -.2273 & -.0957 & -.9691 & -303.688 \\ -.2815 & .9591 & -.0286 & 343.134 \\ 0 & 0 & 0 & 1 \end{bmatrix}$
Calibrated Pose	$\begin{bmatrix} .0780 & .5063 & -.8588 & -269.263 \\ .0419 & .8590 & .5102 & 164.279 \\ .9961 & -.0758 & .0458 & 362.220 \\ 0 & 0 & 0 & 1 \end{bmatrix}$	$\begin{bmatrix} .9322 & .2662 & -.2450 & -73.885 \\ -.2273 & -.0957 & -.9691 & -303.688 \\ -.2815 & .9591 & -.0286 & 343.134 \\ 0 & 0 & 0 & 1 \end{bmatrix}$

L_{r1}, L_{c2}, L_{r2}, and L_{c2}. The last two revolute joint (J_{r2} and J_{r3}) axes are normally parallel with each other. The AIM is given as follows:

$$\mathcal{A}(\tilde{G}) = \begin{bmatrix} (z,-y) & 0 & 0 & 0 & 0 & 0 & L_{r1} \\ (-z,y) & (z,y) & 0 & 0 & 0 & 0 & L_{p1} \\ 0 & (x,-y) & (z,y) & 0 & 0 & 0 & L_{c1} \\ 0 & 0 & (-x,-y) & (z,y) & 0 & 0 & L_{r1} \\ 0 & 0 & 0 & (-x,-y) & (z,y) & 0 & L_{c2} \\ 0 & 0 & 0 & 0 & (y,-x) & (z,x) & L_{r2} \\ 0 & 0 & 0 & 0 & 0 & (z,x) & L_{c2} \\ J_{r1} & J_{p1} & J_{f1} & J_{r2} & J_{f3} & J_{r3} & 0 \end{bmatrix}.$$

According to the given AIM, the calibration model can be automatically generated. Similar to the previous example, we first preset small geometric errors dp_i, ds_i, and dq_i ($i = 1, 2, \ldots 6$) into Dyad i (Table 5.4), and then compute a number of measured end link poses ($m = 40$). For convenience, the dp_i in each of the dyads is preset to the identical value, i.e., dp_i=(2.,2.,2., .02, .02, .02)^T. Since both joint 3 and 5 are fixed joints, ds_3, ds_5, dq_3, and dq_5 are all set to zero. In order to demonstrate the robustness of the proposed algorithm, a uniform random measurement noise dp'=(dx', dy', dz', $\delta x'$, $\delta y'$, $\delta z'$)^T is also added into the measured end link poses, where dx', dy', and $dz' \in$ U[-0.1 mm, 0.1 mm], and $\delta x'$, $\delta y'$, and $\delta z' \in$ U[-0.001, 0.001] rad. The calibration results are shown in Tables 5.5 and 5.6 and Fig. 5.8.

Since joint 3 in Dyad 3 (modules 2 and 3) and joint 5 in Dyad 5 (modules 4 and 5) are fixed joints, the two modules in both Dyad 3 and Dyad 5 are rigidly connected. For simplicity, we consider each of these dyads as a rigid body. The kinematic error in such a dyad will be lumped into its preceding dyad and succeeding dyad. For instance, the preset error dp_3 in Dyad 3 will be lumped into the kinematic errors

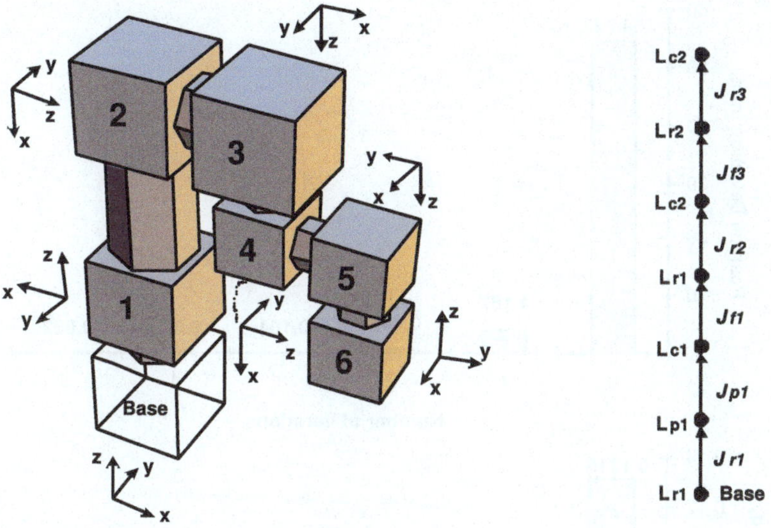

Fig. 5.7 A SCARA type modular robot configuration

Table 5.4 Preset geometric errors for the SCARA robot

	dp_i	ds_i	dq_i
Dyad 1	$(2., 2., 2., 0.02, 0.02, 0.02)^T$	$(0, 2., 0, 0.02, 0, -0.0002)^T$	0.02
Dyad 2	$(2., 2., 2., 0.02, 0.02, 0.02)^T$	$(0.0002, 0.02, 0, 0, 0, 0)^T$	0.02
Dyad 3	$(2., 2., 2., 0.02, 0.02, 0.02)^T$	$(0, 0, 0, 0, 0, 0)^T$	0
Dyad 4	$(2., 2., 2., 0.02, 0.02, 0.02)^T$	$(0, 0, 2., -0.0002, 0.02, 0)^T$	0.02
Dyad 5	$(2., 2., 2., 0.02, 0.02, 0.02)^T$	$(0, 0, 0, 0, 0, 0)^T$	0
Dyad 6	$(2., 2., 2., 0.02, 0.02, 0.02)^T$	$(2., 0, 0, 0, 0.02, 0.0002)^T$	0.02

of both Dyad 2 (modules 1 and 2) and Dyad 4 (modules 3 and 4). Thus, instead of identifying dp_3 and dp_5, we assume $dp_3 = dp_5 = (0, 0, 0, 0, 0, 0)^T$.

The computation results also show that the average positioning accuracy is significantly increased after a few iterations regardless of the existence of the measurement noise. This robot contains various types of joints, i.e., prismatic joint (joint 2), fixed joints (joints 3 and 5), and revolute joints (joints 1, 4, and 6) in which two adjacent ones (joints 4 and 6) are close to parallel. Hence, the proposed algorithm is a general and singular-free approach for the calibration of serial robots.

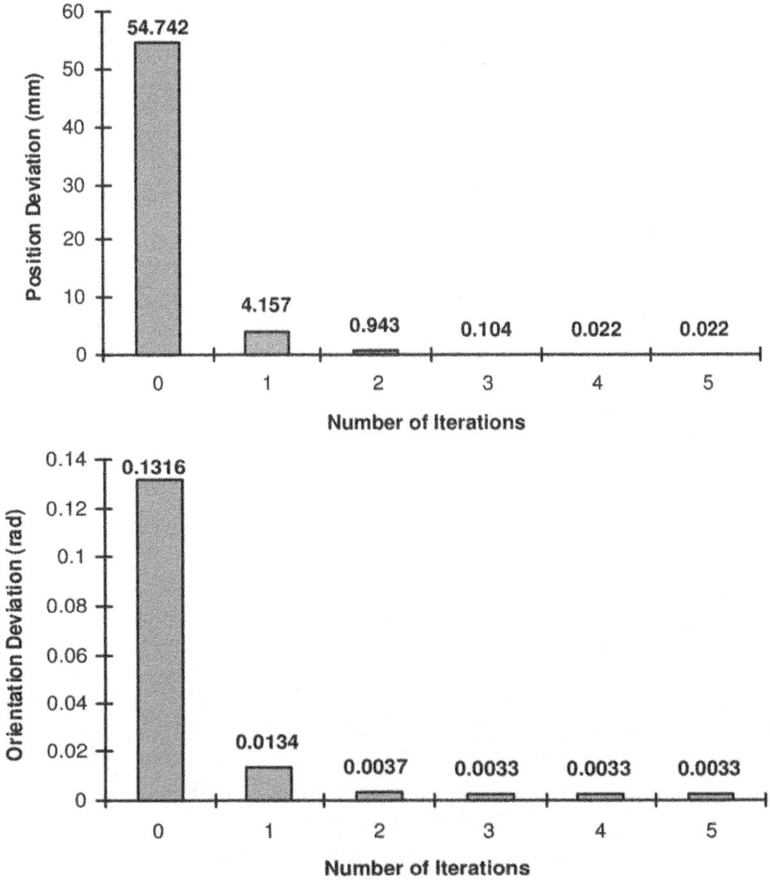

Fig. 5.8 Calibration convergence (SCARA robot)

Table 5.5 Identified kinematic errors for the SCARA robot

	dp_i
Dyad 1	$(-0.110, 2.016, -0.661, 0.0171, 0.0413, -0.1044)^T$
Dyad 2	$(0.628, -0.020, 11.561, -0.1634, 0.0405, -0.0024)^T$
Dyad 3	$(0, 0, 0, 0, 0, 0)^T$
Dyad 4	$(0.675, 0.001, 11.581, 0.1447, 0.0010, 0.0800)^T$
Dyad 5	$(0, 0, 0, 0, 0, 0)^T$
Dyad 6	$(0.001, 3.577, -0.678, 0.0566, -0.0192, 0.0618)^T$
Compensation Dyad	$(0.001, -2.010, 0.741, 0.0201, -0.0005, 0.0338)^T$

Table 5.6 End link poses before and after calibration (SCARA robot)

Joint Angle	$q = (1.3259, .0174, 0, .7298, 0, 2.2093)^T$	$q = (-.2129, 32.721, 0, 1.4841, 0, 1.9221)^T$
Nominal Pose	$\begin{bmatrix} -.9991 & -.0425 & 0 & 312.442 \\ -.0425 & -.9991 & 0 & 493.938 \\ 0 & 0 & 1 & 387.517 \\ 0 & 0 & 0 & 1 \end{bmatrix}$	$\begin{bmatrix} -.0516 & .9987 & 0 & -423.268 \\ -.9987 & -.0516 & 0 & 188.793 \\ 0 & 0 & 1 & 420.221 \\ 0 & 0 & 0 & 1 \end{bmatrix}$
Measured Pose	$\begin{bmatrix} -.9946 & -.1019 & -.0180 & 286.698 \\ .1018 & -.9948 & .0053 & 479.362 \\ -.0184 & .0034 & .9998 & 349.532 \\ 0 & 0 & 0 & 1 \end{bmatrix}$	$\begin{bmatrix} -.1092 & .9907 & -.0808 & -389.917 \\ -.9939 & -.1077 & .0222 & 155.561 \\ .0133 & .0827 & .9965 & 363.005 \\ 0 & 0 & 0 & 1 \end{bmatrix}$
Calibrated Pose	$\begin{bmatrix} -.9947 & -.1020 & -.0151 & 286.686 \\ .1019 & -.9948 & .0068 & 479.356 \\ -.0157 & .0052 & .9999 & 349.509 \\ 0 & 0 & 0 & 1 \end{bmatrix}$	$\begin{bmatrix} -.1094 & .9907 & -.0813 & -389.917 \\ -.9939 & -.1077 & .0254 & 155.555 \\ .0164 & .0836 & .9964 & 363.005 \\ 0 & 0 & 0 & 1 \end{bmatrix}$

Table 5.7 Preset geometric errors for the tree-structured robot

	dp_i	ds_i	dq_i
Dyad 1	$(2., 2., 2., 0.02, 0.02, 0.02)^T$	$(0, 0, 2., 0.02, 0.0002, 0)^T$	0.02
Dyad 2	$(2., 2., 2., 0.02, 0.02, 0.02)^T$	$(0, 0, 2., 0.02, -0.0002, 0)$	0.02
Dyad 3	$(0, 0, 0, 0, 0, 0)^T$	$(0, 0, 0, 0, 0, 0)^T$	0
Dyad 4	$(2., 2., 2., 0.02, 0.02, 0.02)^T$	$(0, 2., 0, 0.0002, 0, 0.02)^T$	0.02
Dyad 5	$(2., 2., 2., 0.02, 0.02, 0.02)^T$	$(0, 0, 2., 0, 0.02, 0.0002, 0)^T$	0.02
Dyad 6	$(2., 2., 2., 0.02, 0.02, 0.02)^T$	$(0, 0.0002, 0.02, 0, 0, 0)^T$	2.
Dyad 7	$(2., 2., 2., 0.02, 0.02, 0.02)^T$	$(0, 0, 2., 0.02, 0.0002, 0)^T$	0.02
Dyad 8	$(2., 2., 2., 0.02, 0.02, 0.02)^T$	$(0, 0, 2., 0.02, 0.0002, 0)^T$	0.02

Table 5.8 Identified kinematic errors for the tree-structured robot

	dp_i
Dyad 1	$(0.078, -0.044, 1.998, 0.0201, 0.0172, 0.0398)^T$
Dyad 2	$(0.574, 1.087, 0.050, 0.0196, 0.0062, 0.0172)^T$
Dyad 3	$(0, 0, 0, 0, 0, 0)^T$
Dyad 4	$(-1.850, 0, 1.084, -0.0249, 0.0400, 0.0343)^T$
Dyad 5	$(7.695, 0.005, 0.572, 0.0590, -0.0306, 0.0605)^T$
Dyad 6	$(-1.852, -0.971, 0, 0.0045, 0.0111, -0.0001)^T$
Dyad 7	$(0, 1.961, 0.003, 0.0206, 0.0277, 0.0467)^T$
Dyad 8	$(0, 0.960, 1.963, 0.0228, -0.0388, 0.0370)^T$
Compensation Dyad (branch 1)	$(1.851, 0.970, -0.004, -0.0197, -0.0088, -0.0002)^T$
Compensation Dyad (branch 2)	$(-1.993, -0.994, -0.005, 0.0016, -0.0586, 0.0204)^T$

Table 5.9 End link poses before and after calibration (tree-structured robot)

	Branch 1 (Module 6)	Branch 2 (Module 8)
Nominal Pose	$\begin{bmatrix} -.8574 & .1383 & -.4957 & 168.263 \\ .0262 & -.9502 & -.3105 & 834.453 \\ -.5140 & -.2792 & .8111 & 646.864 \\ 0 & 0 & 0 & 1 \end{bmatrix}$	$\begin{bmatrix} -.8753 & -.3480 & .3357 & 304.480 \\ .4707 & -.4543 & .7563 & -354.641 \\ -.1107 & .8200 & .5615 & 399.491 \\ 0 & 0 & 0 & 1 \end{bmatrix}$
Measured Pose	$\begin{bmatrix} -.8581 & .1666 & -.4858 & 156.437 \\ .0227 & -.9327 & -.3600 & 814.671 \\ -.5130 & -.3199 & .7965 & 699.125 \\ 0 & 0 & 0 & 1 \end{bmatrix}$	$\begin{bmatrix} -.8892 & -.3799 & .2549 & 295.020 \\ .4299 & -.5032 & .7497 & -343.445 \\ -.1566 & .7762 & .6108 & 406.097 \\ 0 & 0 & 0 & 1 \end{bmatrix}$
Calibration Poses	$\begin{bmatrix} -.8581 & .1667 & -.4857 & 156.451 \\ .0226 & -.9327 & -.3600 & 814.665 \\ -.5130 & -.3199 & .7965 & 699.127 \\ 0 & 0 & 0 & 1 \end{bmatrix}$	$\begin{bmatrix} -.8892 & -.3786 & .2566 & 294.994 \\ .4298 & -.4995 & .7522 & -343.445 \\ -.1566 & .7792 & .6069 & 406.088 \\ 0 & 0 & 0 & 1 \end{bmatrix}$

5.3.3 Calibration of a Tree-structured Robot

In this example, we wish to calibrate a tree-structured modular robot as shown in Fig. 5.4. According to the given AIM and path matrix, the calibration model can also be automatically generated. Similar to previous examples, we first preset small geometric errors dp_i, ds_i, and dq_i $(i = 1, 2, \ldots 8)$ into Dyad i (Table 5.7), and then compute a number of measured poses ($m = 40$) for each of the two end link frames. All dp_i in the dyads are still preset to the identical value, i.e., dp_i=(2.,2.,2., 0.02, 0.02, 0.02)^T. A uniform random measurement noise is also added into the two measured end link poses, which is the same as the previous example. The calibration results are shown in Tables 5.8 and 5.9 and Figs. 5.9 and 5.10. Note that for Table 5.9, the corresponding joint angles are q = (1.5403,1.0310,0,0.3315,-0.6231,82.032,-2.5323,1.3762)^T.

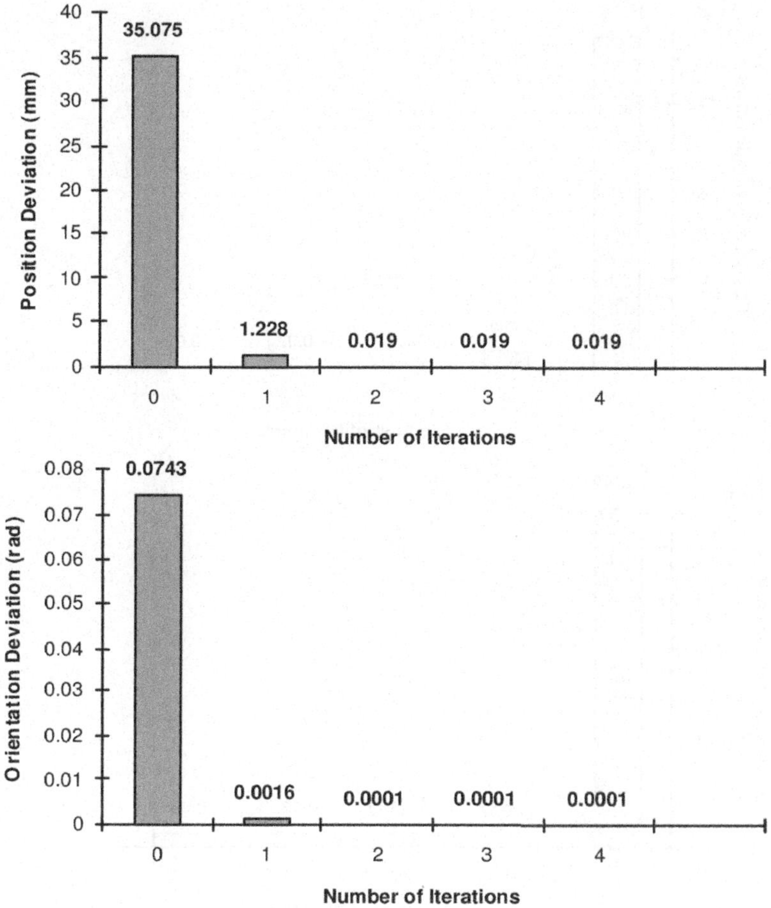

Fig. 5.9 Calibration convergence for branch one

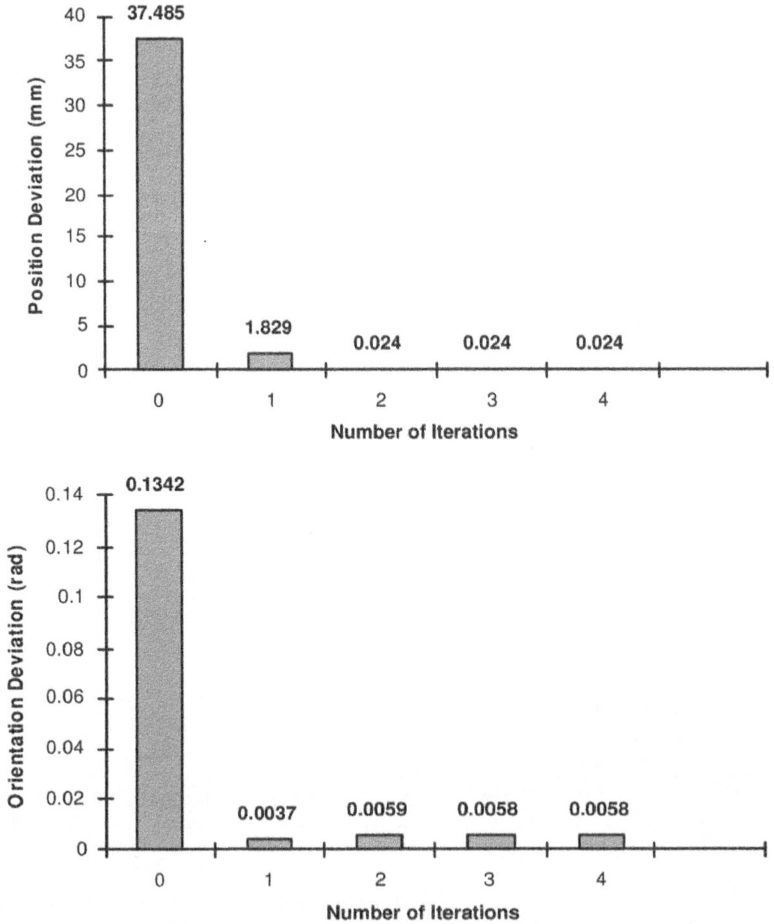

Fig. 5.10 Calibration convergence for branch two

The computation results also show that the average positioning accuracy in each of the two end links is significantly increased after a few iterations regardless of the existence of the measurement noise. This robot has a tree-structured topology and contains various types of joints, i.e., the prismatic joint, fixed joints, and revolute joints. The proposed algorithm therefore is not only suitable for serial robots but also suitable for tree-structured robots.

Chapter 6
Modular Serial Robot Dynamics

Abstract This chapter studies the dynamic issues of modular robots. Since a modular robot has unfixed configurations, it is very difficult to derive the closed-form dynamic model manually. Following the geometric modeling approach, a dynamic formulation scheme is introduced, which can generate both the recursive and the closed-form dynamic models of a tree-structured modular robot from a given Assembly Incidence Matrix (AIM) automatically.

6.1 Introduction

There are two basic issues in modular robot dynamics, i.e., (1) the computation of the actuator torque problem—inverse dynamics, which is to find the actuator forces or torques required to produce a given motion, and (2) the simulation problem—forward dynamics, which is to predict the motion of the robot corresponding to a given time-history of the actuator forces and torques.

Algorithms to solve these two problems can be based on many different general formulations of dynamics such as the Lagrange-Euler (L-E) method [67, 68], the Newton-Euler (N-E) method [115, 116], the Kane's method [117, 118], the recursive Lagrange method [120], and the generalized D'Alembert method [119]. The motion equations in these methods are "equivalent" to each other in the sense that they describe the dynamic behavior of the same physical robot. However, the structure of these equations may differ because they are obtained for various reasons and purposes: to achieve fast computation time in evaluating the nominal joint torques in the robot actuators, to facilitate control analysis and synthesis, and to improve computer simulation of the robot motion. Among these methods, the L-E method and the N-E method have been widely used.

The derivation of the dynamic model of a robot based on the L-E method is simple and systematic. Assuming the robot follows the rigid body motion, the result-

G. Yang and I.-M. Chen, *Modular Robots: Theory and Practice*,
Research on Intelligent Manufacturing,
https://doi.org/10.1007/978-981-16-5007-9_6

ing dynamic equations of motion, excluding the dynamics of the electronic control device, gear friction, and backlash, are a set of second-order coupled, nonlinear, differential equations. Each equation contains a large number of torque or force terms such as inertial torques/forces due to the links, reaction torques/forces generated by acceleration at other joints, velocity-generated (Coriolis and centrifugal) reaction torques/forces between joints, and torques/forces due to gravity loading on the links. These torque/force terms depend on the robot's physical parameters and the external load. To improve the speed of computation, simplified sets of equations have been used in [68]. In general, these "approximate" models simplify the underlying physics by neglecting second-order terms such as the Coriolis and centrifugal reaction terms. These approximate models, when used for control purpose, restrict the arm movement to low speed. At high speed, the neglected terms become significant, making accurate position control of the robot impossible.

An approach that has the advantage of both computation speed and accuracy is based on the N-E vector formulation. The derivation is simple, but messy, and involves vector cross-product terms. The resulting dynamic equations, excluding the dynamics of the electronic control device, gear friction, and backlash, are a set of forward and backward recursive equations. The forward recursion propagates kinematics information (such as angular and linear velocities and accelerations) from the base frame to the end-effector frame, while the backward recursion propagates the forces and moments exerted on each link from the end-effector frame to the base frame. The equations of motion derived from the N-E method represent the most efficient set of equations of motion of the robot, which makes it possible to achieve real-time control. However, the N-E method is difficult to use for deriving advanced control laws because the recursive equations destroy the "structure" of the dynamic model, which provides useful insight for the controller and robot design.

The geometric formulation of the equations of motion for kinematic chains is a new trend in robot dynamics. By treating $SE(3)$, the special Euclidean group of rigid-body motions as Lie group, this geometric approach can avoid many specific definitions and notations found in the conventional dynamic algorithms. Moreover, it permits a coordinate-free view of robot dynamics, which is more suitable for modular robot applications in which the dynamic algorithm is to be configuration independent. So far, a number of fundamental works have been done by several researchers [62, 88, 114, 121–123]. Among these works, a unified geometric treatment of robot dynamics was presented by Park et al. in [88, 114]. In their works, the geometric formulation of the recursive N-E dynamics was presented, and followed by the closed-form Lagrangian formulation. The recursive formulation leads to an $O(n)$ algorithm that expresses the dynamics entirely in terms of the coordinate-free Lie algebraic operations. The Lagrangian formulation also expresses the dynamics in terms of these Lie algebraic operations and leads to a particular simple set of closed-form equations. The resulting equations are shown to be computationally effective and to be easily differentiated and factored with respect to any of the robot parameters [88]. These features make the geometric formulation approach very attractive for applications such as robot design, optimal control, and motion optimization.

Fig. 6.1 Coordinate system in a link assembly

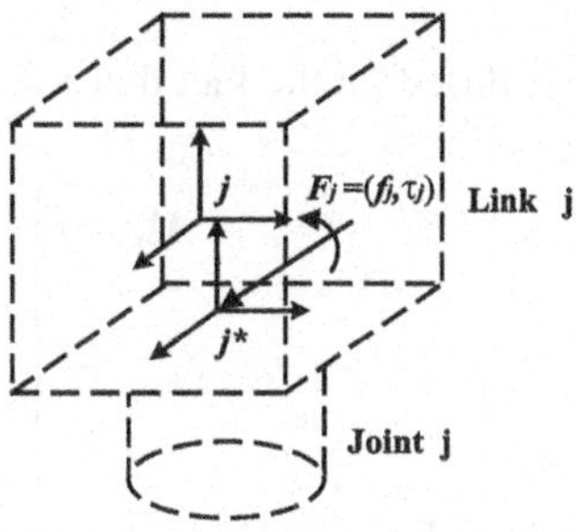

Following the geometrical approach, this chapter presents a methodology that can generate the dynamic model of a modular robot from a given Assembly Incidence Matrix (AIM) automatically [15, 20, 21]. The more general tree-structured robots are considered as they are gaining popularity in industry, construction, and multi-legged walking systems [124]. The formulation of the dynamic model is started with a modified recursive N-E algorithm. The generalized velocities, accelerations, and forces are expressed in terms of linear operations on $se(3)$. Based on the relationship between the recursive formulation and the closed-form Lagrangian formulation for serial robot dynamics discussed in [88, 125, 126], we use an accessibility matrix [127] to assist the construction of the closed-form equations of motion of a tree-structured modular robot.

6.2 Newton-Euler Equation for a Link Assembly

As mentioned in Chap. 4, let v_i and v_j be two adjacent links connected by joint e_j as shown in Fig. 4.1 and v_i precedes v_j. Denote the module coordinate frame on v_i by frame i, v_j by frame j. Frame i and j are module frames located at the centroid of the cubes (Fig. 3.6). Note that joint e_j and link v_j are rigidly connected together, termed link assembly j.

Now consider two coordinate frames located in link assembly j shown in Fig. 6.1. Link frame j is the module (local) coordinate frame defined for the modular robot kinematics. The mass frame j^* is located at the center of mass of the entire link assembly j. Normally, these two frames do not coincide with each other. Let $T_{j^*j} = (p_{j^*j}, R_{j^*j}) \in SE(3)$ be the pose of link frame j relative to mass frame j^*, where $p_{j^*j} \in \Re^{3\times 1}$ and $R_{j^*j} \in SO(3)$ are the position vector and rotation matrix, respectively. Note that T_{j^*j} is a constant because frames j and j^* are located on the same rigid link assembly.

The N-E equation of this rigid link assembly with respect to its mass frame j^* is given by [62]

$$\mathbf{F}_{j^*} = \begin{bmatrix} f_{j^*} \\ \tau_{j^*} \end{bmatrix} = \begin{bmatrix} m_j I & 0 \\ 0 & J_{j^*} \end{bmatrix} \begin{bmatrix} \dot{v}_{j^*} \\ \dot{w}_{j^*} \end{bmatrix} + \begin{bmatrix} w_{j^*} \times m_j v_{j^*} \\ w_{j^*} \times J_{j^*} w_{j^*} \end{bmatrix}. \qquad (6.1)$$

Based on the fact that $v_{j^*} \times m_j v_{j^*} = 0$, this equation can be written as

$$\mathbf{F}_{j^*} = \begin{bmatrix} m_j I & 0 \\ 0 & J_{j^*} \end{bmatrix} \begin{bmatrix} \dot{v}_{j^*} \\ \dot{w}_{j^*} \end{bmatrix} - \begin{bmatrix} -\hat{w}_{j^*} & 0 \\ -\hat{v}_{j^*} & -\hat{w}_{j^*} \end{bmatrix} \begin{bmatrix} m_j I & 0 \\ 0 & J_{j^*} \end{bmatrix} \begin{bmatrix} v_{j^*} \\ w_{j^*} \end{bmatrix}, \tag{6.2}$$

where $\mathbf{F}_{j^*} \in \Re^{6\times1}$ is the resultant wrench applied to the center of mass relative to frame j^*; f_{j^*} and $\tau_{j^*} \in \Re^{3\times1}$ are the total force and moment applied to link assembly j, respectively; $M_{j^*} = \begin{bmatrix} m_j I & 0 \\ 0 & J_{j^*} \end{bmatrix} \in \Re^{6\times6}$ is the generalized mass matrix expressed in frame j^*. The total mass of link assembly j is m_j (the combined mass of link $\mathrm{v_j}$ and joint $\mathrm{e_j}$). J_{j^*} is the inertia tensor of the link assembly about mass frame j^*. Transforming Eq. (6.2) into the adjoint representation, we have

$$\mathbf{F}_{j^*} = M_{j^*}\dot{V}_{j^*} - ad^T_{V_{j^*}}(M_{j^*}V_{j^*}). \tag{6.3}$$

The following notations are adopted in Eq. (6.3):

- $V_{j^*} = \begin{bmatrix} v_{j^*} \\ w_{j^*} \end{bmatrix} \in \Re^{6\times1}$ is the generalized body velocity. v_{j^*} and w_{j^*} are 3×1 vectors defining the body translational velocity, $v_{j^*} = (v_x, v_y, v_z)^T$, and the angular velocity, $w_{j^*} = (w_x, w_y, w_z)^T$, respectively.
- $\hat{v}_{j^*}$ and $\hat{w}_{j^*} \in \Re^{3\times3}$ are skew symmetric matrices related to v_{j^*} and w_{j^*} having the following form:

$$\hat{v}_{j^*} = \begin{bmatrix} 0 & -v_z & v_y \\ v_z & 0 & -v_x \\ -v_y & v_x & 0 \end{bmatrix}; \hat{w}_{j^*} = \begin{bmatrix} 0 & -w_z & w_y \\ w_z & 0 & -w_x \\ -w_y & w_x & 0 \end{bmatrix}.$$

- $ad^T_{V_{j^*}} \in \Re^{6\times6}$ is the transpose of the adjoint matrix $ad_{V_{j^*}}$ related to V_{j^*} such that

$$ad^T_{V_{j^*}} = (ad_{V_{j^*}})^T = \begin{bmatrix} \hat{w}_{j^*} & \hat{v}_{j^*} \\ 0 & \hat{w}_{j^*} \end{bmatrix}^T = \begin{bmatrix} -\hat{w}_{j^*} & 0 \\ -\hat{v}_{j^*} & -\hat{w}_{j^*} \end{bmatrix}. \tag{6.4}$$

- $\dot{V}_{j^*} = \begin{bmatrix} \dot{v}_{j^*} \\ \dot{w}_{j^*} \end{bmatrix} \in \Re^{6\times1}$ is the generalized body acceleration.

The N-E equation, Eq. (6.3), is expressed in mass frame j^*. To be consistent with the kinematic model of modular robots described in Chap. 4, this equation must be expressed in the module frame j. According to the wrench transformation method mentioned in [62], the wrench $\mathbf{F}_{j^*}$ can be transformed into an equivalent wrench $\mathbf{F}_j$ expressed in link frame j by

$$\mathbf{F}_j = Ad^T_{T_{j^*,j}}\mathbf{F}_{j^*}, \tag{6.5}$$

where $Ad^T_{T_{j^*,j}}$ is the transpose of $Ad_{T_{j^*,j}}$ such that

$$
\begin{aligned}
Ad^T_{T_{j^*,j}} &= (Ad_{T_{j^*,j}})^T = \begin{bmatrix} R_{j^*j} & \hat{p}_{j^*j}R_{j^*j} \\ 0 & R_{j^*j} \end{bmatrix}^T \\
&= \begin{bmatrix} R^T_{j^*j} & 0 \\ -R^T_{j^*j}\hat{p}_{j^*j} & R^T_{j^*j} \end{bmatrix}.
\end{aligned}
\tag{6.6}
$$

Note that $\hat{p}_{j^*j}$ is the 3×3 skew symmetric matrix representation of p_{j^*j}. Similarly, the generalized body velocity V_{j^*} and the generalized body acceleration $\dot{V}_{j^*}$ expressed in frame j^* can also be transformed into the equivalent V_j and $\dot{V}_j$ in link frame j as follows:

$$V_{j^*} = Ad_{T_{j^*,j}} V_j, \tag{6.7}$$

$$\dot{V}_{j^*} = Ad_{T_{j^*,j}} \dot{V}_j. \tag{6.8}$$

Substituting Eqs. (6.5), (6.7), and (6.8) into (6.3), and using the identity,

$$Ad_{T_{j^*,j}} ad_{V_j} = ad_{Ad_{T_{j^*,j}}V_j} Ad_{T_{j^*,j}}, \tag{6.9}$$

we have

$$\mathbf{F}_j = M_j \dot{V}_j - ad^T_{V_j}(M_j V_j). \tag{6.10}$$

This is the N-E equation of link assembly j written in the link frame j. The generalized mass matrix M_j is relative to link frame j and has the following form:

$$
\begin{aligned}
M_j &= Ad^T_{T_{j^*,j}} M_{j^*} Ad_{T_{j^*,j}} \\
&= \begin{bmatrix} m_j I & m_j R^T_{j^*j}\hat{p}_{j^*j}R_{j^*j} \\ -m_j R^T_{j^*j}\hat{p}_{j^*j}R_{j^*j} & R^T_{j^*j}(J_{j^*} - m_j \hat{p}^2_{j^*j})R_{j^*j} \end{bmatrix}.
\end{aligned}
\tag{6.11}
$$

For simplicity, we assume that the coordinate axes of mass frame j^* are parallel to the axes of link frame j, then $R_{j^*j} = I$. The generalized mass matrix M_j becomes

$$M_j = \begin{bmatrix} m_j I & m_j \hat{p}_{j^*j} \\ -m_j \hat{p}_{j^*j} & J_{j^*} - m_j \hat{p}^2_{j^*j} \end{bmatrix}. \tag{6.12}$$

Equations (6.3) and (6.10) are of the same form indicating the coordinate-invariant property of the N-E equation in body coordinates.

6.3 Dynamic Formulation for a Tree-Structured Modular Robot

6.3.1 Recursive Newton-Euler Algorithm

The recursive algorithm is a two-step iteration process. For a tree-structured robot, the generalized velocity and acceleration of each link are propagated from the base to the end links of all branches. The generalized force of each link is propagated backward from the end links of all branches to the base. At the branching module, generalized forces transmitted back from all branches are combined.

Forward iteration

The general velocity and acceleration of the base link are given initially:

$$V_b = V_0 = (0, 0, 0, 0, 0, 0)^T, \tag{6.13}$$
$$\dot{V}_b = \dot{V}_0 = (0, 0, g, 0, 0, 0)^T, \tag{6.14}$$

where V_b and $\dot{V}_b$ are expressed in the base frame 0. We assume that the base frame coincides with the world frame. The gravity effect is taken into consideration by initializing the generalized acceleration, Eq. (6.14), with the gravitational acceleration constant g. The recursive body velocity and acceleration equations can be written as

$$V_j = Ad_{T_{i,j}^{-1}}(V_i) + s_j\dot{q}_j, \tag{6.15}$$
$$\dot{V}_j = Ad_{T_{i,j}^{-1}}(\dot{V}_i) + ad_{Ad_{T_{i,j}^{-1}}(V_i)}(s_j\dot{q}_j) + s_j\ddot{q}_j, \tag{6.16}$$

where all the quantities, if not specified, are expressed in link frame j.

- V_j and $\dot{V}_j$ are the generalized velocity and acceleration of link assembly j.
- V_i and $\dot{V}_i$ are the generalized velocity and acceleration of link assembly i written in frame i.
- $\dot{q}_j$ and $\ddot{q}_j$ are the velocity and acceleration of joint e_j, respectively.
- $Ad_{T_{i,j}^{-1}}$ is the adjoint representation of $T_{i,j}^{-1}(q_j)$, where $T_{i,j}(q_j) \in SE(3)$ is the pose of frame j relative to frame i with the joint displacement q_j.
- $s_j \in \Re^{6\times 1}$ is the twist coordinates of joint e_j, and $s_j = (0, 0, 0, 0, 0, 0)^T$ for a fixed joint.

Both $Ad_{T_{i,j}^{-1}}(V_i)$ and $s_j\dot{q}_j$ are elements of $se(3)$, and

$$ad_{Ad_{T_{i,j}^{-1}}(V_i)}(s_j\dot{q}_j) = -ad_{s_j\dot{q}_j}(Ad_{T_{i,j}^{-1}}(V_i)). \tag{6.17}$$

In Eq. (6.15), we find that

$$Ad_{T_{i,j}^{-1}}(V_i) = V_j - s_j\dot{q}_j. \tag{6.18}$$

Substituting Eq. (6.18) into (6.17), we have

$$ad_{Ad_{T_{i,j}^{-1}}(V_i)}(s_j\dot{q}_j) = -ad_{s_j\dot{q}_j}(V_j). \tag{6.19}$$

Therefore, Eq. (6.16) can also be written as

$$\dot{V}_j = Ad_{T_{i,j}^{-1}}(\dot{V}_i) - ad_{s_j\dot{q}_j}(V_j) + s_j\ddot{q}_j. \tag{6.20}$$

Backward Iteration

The backward iteration of a tree-structured robot starts from all the pendant link assemblies simultaneously. Let $\mathcal{V}_{PD}$ be set of the pendant links of a tree-structured robot. For every pendant link assembly d_i ($\mathrm{v}_{d_i} \in \mathcal{V}_{PD}$), Eq. (6.10) can be written as

$$F_{d_i} = -F_{d_i}^e + M_{d_i}\dot{V}_{d_i} - ad_{V_{d_i}}^T(M_{d_i}V_{d_i}), \tag{6.21}$$

where F_{d_i} is the wrench exerted on link assembly d_i by its parent (preceding) link relative to frame d_i; $F_{d_i}^e$ is the external wrench exerted on link assembly d_i. Note that the total wrench $\mathbf{F}_{d_i} = F_{d_i} + F_{d_i}^e$.

Now traverse the link assemblies in the tree-structured robot backward from the pendant link assemblies. Let $\mathcal{V}_{Hi}$ be the set of successors of link v_i as viewed from the base to the pendant link assemblies. For every link assembly i, Eq. (6.10) based on the N-E method can be written in the following form:

$$F_i = \sum_{j\in V_{Hi}} Ad_{T_{i,j}^{-1}}^T(F_j) - F_i^e + M_i\dot{V}_i - ad_{V_i}^T(M_iV_i), \tag{6.22}$$

where all quantities, if not specified, are expressed in link frame i; $F_i \in \Re^{6\times1}$ is the wrench exerted to link assembly i by its ancestor; $F_j \in \Re^{6\times1}$ is the wrench exerted by link assembly i to the successor $\mathrm{v}_j (\in \mathcal{V}_{Hi})$ expressed in link frame j; F_i^e is the external wrench applied to link assembly i. Basically, the total wrench of link assembly i can be written as: $\mathbf{F}_i = F_i - \sum_{j\in V_{Hi}} Ad_{T_{i,j}^{-1}}^T(F_j) + F_i^e$.

The applied torque/force to link assembly i by the actuator at its input joint $\mathrm{e_i}$ can be calculated by

$$\tau_i = s_i^T F_i. \tag{6.23}$$

6.3.2 Closed Form Equations of Motion

By expanding the recursive N-E Eqs. (6.13), (6.14), (6.15), (6.20), (6.21), and (6.22) in the module (local) frame coordinates iteratively, we obtain the generalized velocity, generalized acceleration, and generalized force equations in matrix form:

$$\mathbf{V} = \mathbf{H}\mathbf{S}\dot{\mathbf{q}}, \tag{6.24}$$
$$\dot{\mathbf{V}} = \mathbf{H}_{\mathbf{T_0}}\dot{\mathbf{V}}_0 + \mathbf{H}\mathbf{S}\ddot{\mathbf{q}} + \mathbf{H}\mathbf{A_1}\mathbf{V}, \tag{6.25}$$
$$\mathbf{F} = \mathbf{H}^T\mathbf{F}^{\mathbf{E}} + \mathbf{H}^T\mathbf{M}\dot{\mathbf{V}} + \mathbf{H}^T\mathbf{A_2}\mathbf{M}\mathbf{V}, \tag{6.26}$$
$$\tau = \mathbf{S}^T\mathbf{F}. \tag{6.27}$$

where
$\mathbf{V} = column[V_1, V_2, \cdots, V_n] \in \Re^{6n\times 1}$ is the generalized body velocity vector;
$\dot{\mathbf{V}} = column[\dot{V}_1, \dot{V}_2, \cdots, \dot{V}_n] \in \Re^{6n\times 1}$ is the generalized body acceleration vector;
$\mathbf{F} = column[F_1, F_2, \cdots, F_n] \in \Re^{6n\times 1}$is the body wrench vector;
$\tau = column[\tau_1, \tau_2, \cdots, \tau_n] \in \Re^{n\times 1}$ is the applied joint torque/force vector;
$\dot{\mathbf{q}} = column[\dot{q}_1, \dot{q}_2, \cdots, \dot{q}_n] \in \Re^{n\times 1}$ is the joint velocity vector;
$\ddot{\mathbf{q}} = column[\ddot{q}_1, \ddot{q}_2, \cdots, \ddot{q}_n] \in \Re^{n\times 1}$ is the joint acceleration vector;
$\dot{\mathbf{V}}_\mathbf{0} = (0, 0, g, 0, 0, 0)^T \in \Re^{6\times 1}$ is the generalized acceleration of the base link;
$\mathbf{S} = diag[s_1, s_2, \cdots, s_n] \in \Re^{6n\times n}$ is the joint twist matrix written in the respective body coordinates;
$\mathbf{M} = diag[M_1, M_2, \cdots, M_n] \in \Re^{6n\times 6n}$ is the total generalized mass matrix;
$\mathbf{A_1} = diag[-ad_{s_1\dot{q}_1}, -ad_{s_2\dot{q}_2}, \cdots, -ad_{s_n\dot{q}_n}] \in \Re^{6n\times 6n}$;
$\mathbf{A_2} = diag[-ad^T_{V_1}, -ad^T_{V_2}, \cdots, -ad^T_{V_n}] \in \Re^{6n\times 6n}$;
$\mathbf{F^E} = column[F^e_1, F^e_2, \cdots, F^e_n] \in \Re^{6n\times 1}$ is the external wrench vector;

$$\mathbf{H}_{\mathbf{T_0}} = \begin{bmatrix} Ad_{T_{01}^{-1}} \\ Ad_{T_{02}^{-1}} \\ \vdots \\ Ad_{T_{0n}^{-1}} \end{bmatrix} \in \Re^{6n\times 6};$$

$$\mathbf{H} = \begin{bmatrix} I_{6\times 6} & 0 & 0 & \cdots & 0 \\ r_{12}Ad_{T_{12}^{-1}} & I_{6\times 6} & 0 & \cdots & 0 \\ r_{13}Ad_{T_{13}^{-1}} & r_{23}Ad_{T_{23}^{-1}} & I_{6\times 6} & \cdots & 0 \\ \vdots & \vdots & \vdots & \ddots & \vdots \\ r_{1n}Ad_{T_{1n}^{-1}} & r_{2n}Ad_{T_{2n}^{-1}} & r_{3n}Ad_{T_{3n}^{-1}} & \cdots & I_{6\times 6} \end{bmatrix} \in \Re^{6n\times 6n}.$$

Note that $\mathcal{R}(\mathbf{G}) = [r_{i,j}] \in \Re^{(n+1)\times(n+1)}$ is the accessibility matrix of the robot's kinematic digraph **G**. The matrix **H** is termed a transmission matrix. The element $r_{i,j}Ad_{T_{i,j}^{-1}}$ relates link frames i and j if link $\mathrm{v_i}$ can reach link $\mathrm{v_j}$. Substituting Eqs. (6.24–6.26) to (6.27), we obtain the closed-form equations of motion for a tree-structured modular robot with $n + 1$ modules (including the base module)

$$\mathbf{M}(\mathbf{q})\ddot{\mathbf{q}} + \mathbf{C}(\mathbf{q}, \dot{\mathbf{q}})\dot{\mathbf{q}} + \mathbf{N}(\mathbf{q}) = \tau, \tag{6.28}$$

where

Table 6.1 Computational complexity

	Multiplications	Additions
Recursive Algorithm	$342n - 103$	$266n - 174$
Closed-form Algorithm	$48n^3 + 68n^2 + 387n - 108$	$48n^3 + 50n^2 + 317n - 78$

$$\mathbf{M}(\mathbf{q}) = \mathbf{S}^T\mathbf{H}^T\mathbf{M}\mathbf{H}\mathbf{S}; \tag{6.29}$$

$$\mathbf{C}(\mathbf{q}, \dot{\mathbf{q}}) = \mathbf{S}^T\mathbf{H}^T(\mathbf{M}\mathbf{H}\mathbf{A}_1 + \mathbf{A}_2\mathbf{M})\mathbf{H}\mathbf{S}; \tag{6.30}$$

$$\mathbf{N}(\mathbf{q}) = \mathbf{S}^T\mathbf{H}^T\mathbf{M}\mathbf{H}_{\mathbf{T}_0}\dot{\mathbf{V}}_0 + \mathbf{S}^T\mathbf{H}^T\mathbf{F}^{\mathbf{E}}. \tag{6.31}$$

$\mathbf{M}(\mathbf{q})$ is the generalized mass matrix of the robot; $\mathbf{C}(\mathbf{q}, \dot{\mathbf{q}})$ is called Coriolis matrix for the robot [62]; $\mathbf{C}(\mathbf{q}, \dot{\mathbf{q}})\dot{\mathbf{q}}$ gives the Coriolis and centrifugal force terms in the equations of motion; $\mathbf{N}(\mathbf{q})$ represents the gravitational force and the external force which act at the joints; τ is the vector of actuator torques/forces.

6.3.3 Remarks on the Dynamics Algorithms

Based on the geometric formulation of the N-E dynamics equations, both of the recursive and closed-form dynamics algorithms have been presented for general tree-structured modular robots. The major advantage of this approach is that the high level and coordinate-invariant description of robot dynamics can be obtained [126], which facilities our automatic model generation scheme. Regarding the computational complexity (as listed in Table 6.1), the recursive geometric N-E model is equivalent to a set of efficient $O(n)$ recursive algorithms, while the closed-form dynamic equations of motion are equivalent to a set of $O(n^3)$ algorithms. Note that it is very difficult to estimate the computational complexity of the dynamics algorithms for a tree-structured robot because it depends on not only the number of DOFs of the robot but also the number of branches and the number of modules on each branch. Hence, we use an n-DOF serial type modular robot to estimate the computational complexity of the derived dynamics algorithms as listed in Table 6.1.

6.3.4 Implementation and Examples

Now, we concentrate on how to generate the closed-form equations of motion automatically, i.e., Eq. (6.28), from a given AIM. To formulate Eq. (6.28), we need to determine the contents of the matrices $\mathbf{M}(\mathbf{q})$, $\mathbf{C}(\mathbf{q}, \dot{\mathbf{q}})$ and $\mathbf{N}(\mathbf{q})$.

1. The joint twists in $\mathbf{S}$ are represented in local module frames and are usually parallel to the coordinate axes. They can be obtained from the AIM directly and remain unchanged during the robot motion.

Fig. 6.2 Flow chart of dynamic model generation

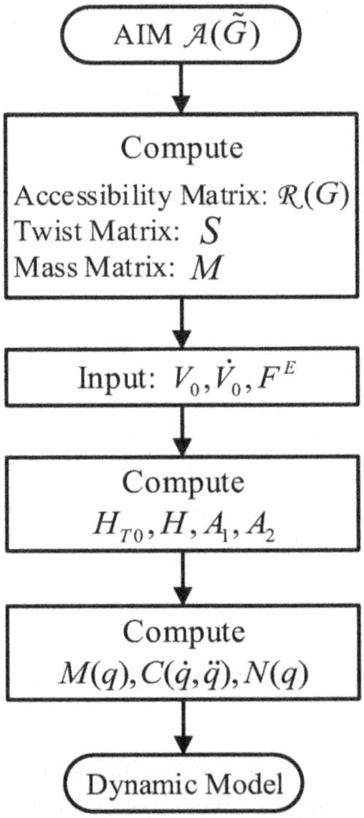

2. The inertia properties of robot modules can be calculated or measured individually, which are listed in Chap. 2. Since the accessibility and path matrices can be derived from the AIM, the generalized mass matrix M_{j*} (relative to the mass frame j^*) and M_j (relative to the module frame j) of link assembly j can be derived automatically.
3. The forward kinematic algorithms proposed in Chap. 4 and [4] can determine the forward kinematic transformations between any two connective modules in a tree-structured modular robot automatically once the AIM and joint angles are given. Hence, all the adjoint representations in $\mathbf{H}$ and $\mathbf{H}_{\mathbf{T}_0}$ can be calculated. The procedure to obtain the closed-form equations of motion for a modular robot is shown in Fig. 6.2.

The dynamic formulation algorithm has been successfully implemented in *Mathematica* code as shown in Fig. 6.2. Once a new modular robot assembly is constructed, a new AIM will be formed and so will the equations of motion of the robot. Note that the symbolic expression of the dynamic equations can be obtained as well. Here we give a computation example to illustrate the proposed algorithm.

Example 6.1 As shown in Fig. 6.3, the modular robot consists of 8 modules and has 2 branches. Its AIM is given as follows:

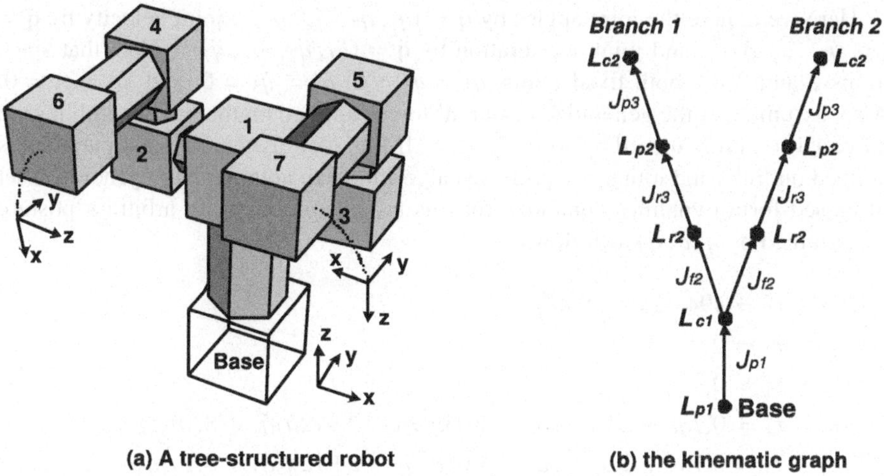

Fig. 6.3 A tree-structured modular robot configuration (5-DOF)

$$
\mathcal{A}(\tilde{G}) = \begin{bmatrix}
(z,-y) & 0 & 0 & 0 & 0 & 0 & 0 & L_{p1} \\
(-z,x) & (-y,x) & (y,x) & 0 & 0 & 0 & 0 & L_{c1} \\
0 & (y,x) & 0 & (z,x) & 0 & 0 & 0 & L_{r2} \\
0 & 0 & (x,y) & 0 & (z,y) & 0 & 0 & L_{r2} \\
0 & 0 & 0 & (x,z) & 0 & (z,x) & 0 & L_{p2} \\
0 & 0 & 0 & 0 & (y,z) & 0 & (z,x) & L_{p2} \\
0 & 0 & 0 & 0 & 0 & (y,x) & 0 & L_{c2} \\
0 & 0 & 0 & 0 & 0 & 0 & (y,x) & L_{c2} \\
J_{p1} & J_{f2} & J_{f2} & J_{r3} & J_{r3} & J_{p3} & J_{p3} & 0
\end{bmatrix}.
$$

Based on the given AIM, the accessibility and path matrices of the kinematic graph shown in Fig. 6.3 can be given by

$$
\mathcal{R}(\mathbf{G}) = \begin{bmatrix}
0\,1\,1\,1\,1\,1\,1\,1 \\
0\,0\,1\,1\,1\,1\,1\,1 \\
0\,0\,0\,0\,1\,0\,1\,0 \\
0\,0\,0\,0\,0\,1\,0\,1 \\
0\,0\,0\,0\,0\,0\,1\,0 \\
0\,0\,0\,0\,0\,0\,0\,1 \\
0\,0\,0\,0\,0\,0\,0\,0 \\
0\,0\,0\,0\,0\,0\,0\,0
\end{bmatrix};
$$

$$
\mathcal{P}(\mathbf{G}) = \begin{bmatrix}
1\,1\,1\,0\,1\,0\,1\,0 \\
0\,1\,0\,1\,0\,1\,0\,1
\end{bmatrix}.
$$

Here we denote the joint angles by $\mathbf{q} = (q_1, q_2, \ldots, q_7)^T$, joint velocity by $\dot{\mathbf{q}} = (\dot{q}_1, \dot{q}_2, \ldots, \dot{q}_7)^T$, and joint acceleration by $\ddot{\mathbf{q}} = (\ddot{q}_1, \ddot{q}_2, \ldots, \ddot{q}_7)^T$. Note that since joints 2 and 3 are both fixed joints, $q_2 = q_3 = 0$; $\dot{q}_2 = \dot{q}_3 = 0$; and $\ddot{q}_2 = \ddot{q}_3 = 0$. Also, assume that the generalized external forces applied on the link assemblies are all equal to zero. According to the given AIM as well as the physical parameters defined on links and joints, the proposed algorithm can automatically generate a set of closed-form dynamics equations for this modular robot at an arbitrary posture (determined by $q, \dot{q}, \ddot{q}$) as follows:

$$
\begin{aligned}
\tau_1 &= 494.424 + 50.4\ddot{q}_1, \\
\tau_2 &= 0, \\
\tau_3 &= 0, \\
\tau_4 &= 0.7\ddot{q}_4 + 2.912q_6\ddot{q}_4 + 4.05q_6^2\ddot{q}_4 + 2.912\dot{q}_4\dot{q}_6 + 8.1q_6\dot{q}_4\dot{q}_6, \\
\tau_5 &= 0.7\ddot{q}_5 + 2.912q_7\ddot{q}_5 + 4.05q_7^2\ddot{q}_5 + 2.912\dot{q}_5\dot{q}_7 + 8.1q_7\dot{q}_5\dot{q}_7, \\
\tau_6 &= 4.05\ddot{q}_6 - 1.456\dot{q}_4^2 - 4.05q_6\dot{q}_4^2, \\
\tau_7 &= 4.05\ddot{q}_7 - 1.456\dot{q}_5^2 - 4.05q_7\dot{q}_5^2.
\end{aligned}
$$

Because the given robot configuration is not very complicated, we can still manually formulate its dynamic model based on the traditional N-E method. It has shown that the manually derived dynamic equations are identical to the equations listed above. Note that SI units are used in the derived equations.

6.4 Inverse and Forward Dynamics Problem

6.4.1 Inverse Dynamics

The inverse dynamics problem is to determine the joint torques $\tau(t)$ which are needed to generate the motion specified by the joint accelerations $\ddot{\mathbf{q}}(t)$, velocities $\dot{\mathbf{q}}(t)$, and angles $\mathbf{q}(t)$ when the external forces $\mathbf{F}^E$ are known.

Solving the inverse dynamics problem is necessary for robot motion planning and control algorithm implementation. Once a robot trajectory is specified in terms of joint angles, velocities, and accelerations, and the external forces are also given, the inverse dynamics allows the computation of the torques to be applied to the joints to obtain the desired motion. This computation turns out to be useful for verifying the feasibility of the imposed trajectory and for compensating nonlinear terms in the dynamic model of a robot. Obviously, the computation of inverse dynamics becomes quite straightforward once the dynamic model is derived.

6.4.2 Forward Dynamics

The forward dynamics problem is to determine the joint accelerations $\ddot{\mathbf{q}}(t)$, velocities $\dot{\mathbf{q}}(t)$, and angles $\mathbf{q}(t)$ resulting from the given joint torques/forces τ and the external forces $\mathbf{F}^E$ when the initial joint angles $\mathbf{q}(t_0)$ and velocities $\dot{\mathbf{q}}(t_0)$ are known.

Solving the forward dynamics problem is necessary for the robot simulation. Forward dynamics allows describing the motion of the real physical system in terms of the joint accelerations, when a set of assigned joint torques is applied to the robot. The joint velocities and angles can be obtained by integrating the system of nonlinear differential equations. The closed-form equations of motion derived from the proposed algorithm are more convenient for this purpose because they provide the explicit relationship between the joint torques (as well as the external forces), angles, velocities, and accelerations. Equation (6.28) can also be rewritten as

$$\ddot{\mathbf{q}} = \mathbf{M}^{-1}(\mathbf{q})(\tau - \mathbf{C}(\mathbf{q}, \dot{\mathbf{q}})\dot{\mathbf{q}} - \mathbf{N}(\mathbf{q})). \tag{6.32}$$

Equation (6.32) consists of a set of second order differential equations. For the simulation of robot motion, the initial state at the time instant t_0 is known with the joint angles $\mathbf{q}(t_0)$ and velocities $\dot{\mathbf{q}}(t_0)$. The forward dynamic problem then solves a set of second order differential equations subjected to the given initial conditions. Many numerical integration methods can be used to derive the solutions. In our forward dynamic algorithm, the fourth order Runge-Kutta method is employed for solutions. The inputs of this algorithm include a given AIM, the initial time instant t_0, the ending time instant t_n, the integration step δt, the initial joint angles $\mathbf{q}(t_0)$, the joint velocities $\dot{\mathbf{q}}(t_0)$, the applied joint torques $\tau(t)$, and the external forces exerted on the each of the link assemblies $\mathbf{F}^{\mathrm{E}}(t)$, where $t_0 \leq t \leq t_n$. The outputs are the joint accelerations, velocities, and angles at different time instants. The flow chart of the forward kinematic algorithm is shown in Fig. 6.4. In order to illustrate this algorithm, a computation example is given as follows.

Example 6.2 In this example, we still use the same robot configuration as well as the AIM as given in Example 6.1. Suppose that $\mathbf{q}(t_0)$, $\dot{\mathbf{q}}(t_0)$, and $\mathbf{F}^{\mathrm{E}}(t)$ are all equal to zero; $\tau(t)$ takes a constant value such that $\tau(t)$=(500, 0, 0, -0.5, 0.5, 0.05, 0.05)^T; and $t_0 = 0$, $t_n = 1.5$s, $\delta t = 0.05$s. Selected computation results are given in Tables 6.2, 6.3, and 6.4, and Figs. 6.5, 6.6, and 6.7. The SI units are used in the computation results. The corresponding robot configurations at different time instants are shown in Fig. 6.8.

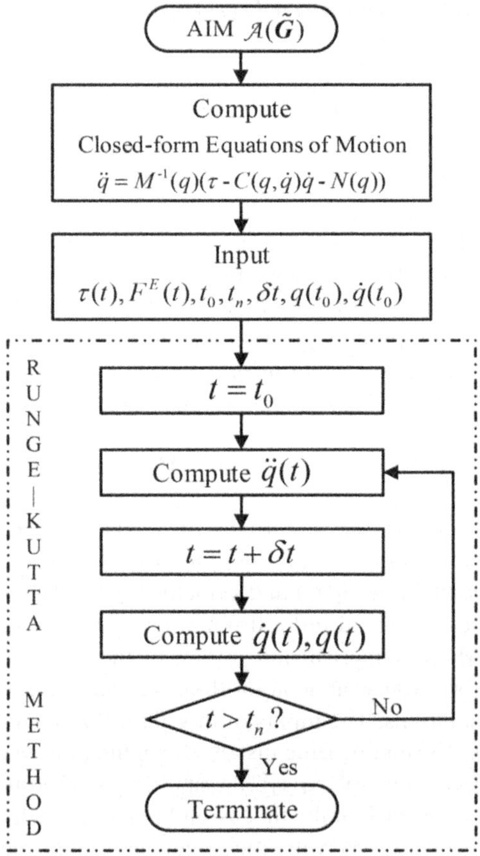

Fig. 6.4 Flow chart of the forward dynamic algorithm

Table 6.2 Joint acceleration ($1/s^2$ or m/s^2)

Time (s)	Joint 1	Joint 2	Joint 3	Joint 4	Joint 5	Joint 6	Joint 7
0	0.1106	0	0	−0.7147	0.7147	0.0123	0.0123
0.3	0.1106	0	0	−0.7080	0.7080	0.0288	0.0288
0.6	0.1106	0	0	−0.6667	0.6667	0.0769	0.0769
0.9	0.1106	0	0	−0.5392	0.5392	0.1494	0.1494
1.2	0.1106	0	0	−0.2913	0.2913	0.2261	0.2261
1.5	0.1106	0	0	0.0193	−0.0193	0.2771	0.2771

Table 6.3 Joint velocity (1/s or m/s)

Time (s)	Joint 1	Joint 2	Joint 3	Joint 4	Joint 5	Joint 6	Joint 7
0	0	0	0	0	0	0	0
0.3	0.0332	0	0	−0.2138	0.2138	0.0054	0.0054
0.6	0.0663	0	0	−0.4215	0.4215	0.0205	0.0205
0.9	0.0996	0	0	−0.6052	0.6052	0.0540	0.0540
1.2	0.1328	0	0	−0.7326	0.7326	0.1106	0.1106
1.5	0.1660	0	0	−0.7734	0.7734	0.1871	0.1871

Table 6.4 Joint displacement (rad or m)

Time (s)	Joint 1	Joint 2	Joint 3	Joint 4	Joint 5	Joint 6	Joint 7
0	0	0	0	0	0	0	0
0.3	0.0050	0	0	−0.0321	0.0321	0.0007	0.0007
0.6	0.0199	0	0	−0.1277	0.1277	0.0042	0.0042
0.9	0.0448	0	0	−0.2827	0.2827	0.0148	0.0148
1.2	0.0797	0	0	−0.4852	0.4852	0.0389	0.0389
1.5	0.1245	0	0	−0.7134	0.7134	0.0832	0.0832

Fig. 6.5 Joint acceleration

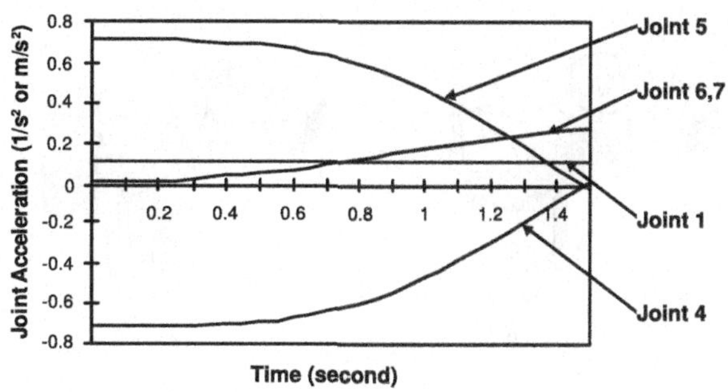

Fig. 6.6 Joint velocity

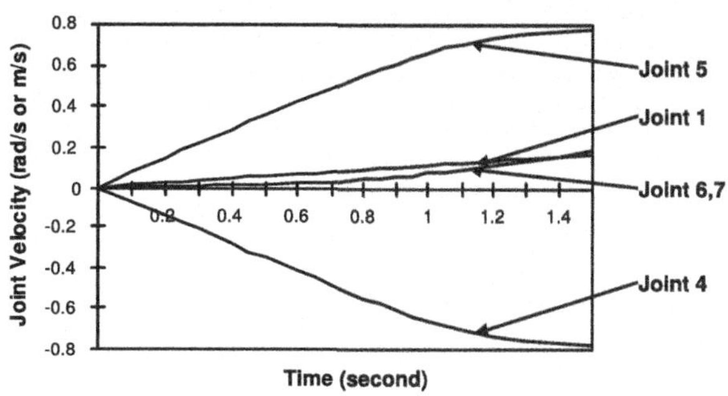

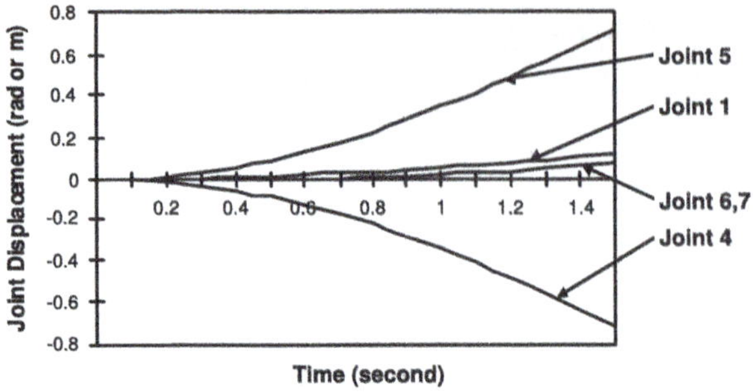

Fig. 6.7 Joint displacement

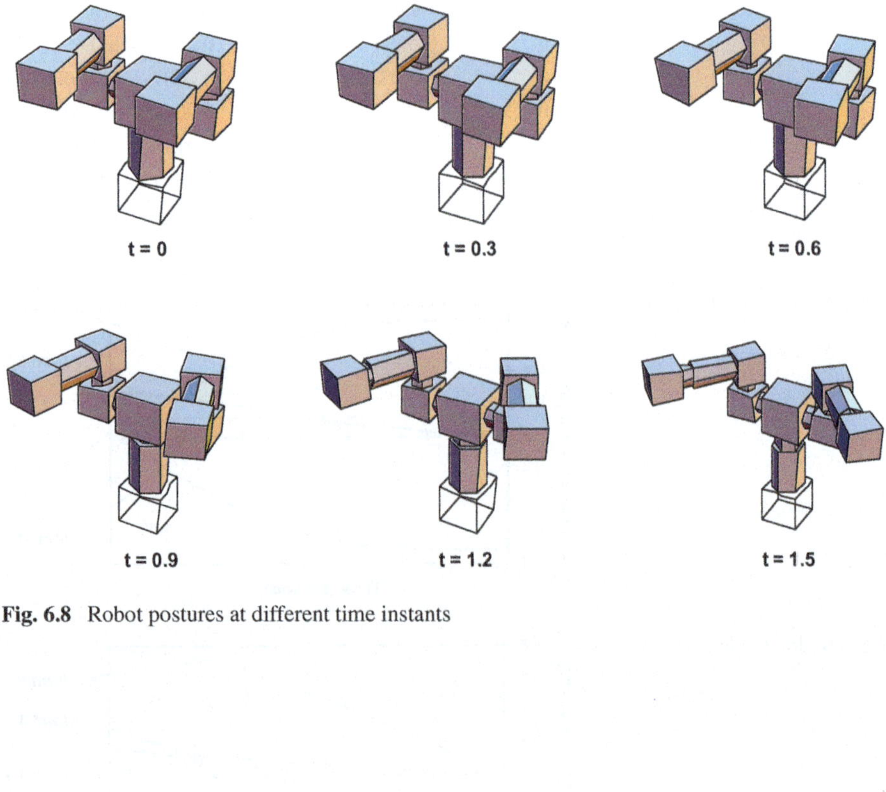

Fig. 6.8 Robot postures at different time instants

Chapter 7
Optimization of Modular Serial Robot Configurations

Abstract This chapter discusses issues pertinent to the optimization of modular robot configurations. The optimization model dealing with the issues of the task specification, the design parameters, the searching space, the objective function, and performance constraints is discussed. Based on the classical Genetic Algorithm (GA) and effective coding schemes defined on Assembly Incidence Matrices (AIMs), an Evolutionary Algorithm (EA) approach is employed to search for optimal solutions. A virtual module concept is introduced into the coding schemes to guarantee the operation of EA. Examples that demonstrate the problem solving strategy are presented.

7.1 Introduction

There has been an increasing interest in studying the kinematic properties of robots recently. Most of these efforts focus on identifying and quantifying certain qualitative features of the robot performance—performance measures. These measures are then used to evaluate and optimize robot configurations. Among these performance measures, the workspace volume and manipulability (dexterity) have been studied extensively in the literatures.

The workspace of a robot is defined as the set of all end-effector poses (described by the end-effector frame) which can be reached by the robot through some choices of joint angles. The relationship between a robot's kinematic parameters and its workspace was first addressed in the early work of Roth [130]. Now it is often customary to distinguish the workspace into the reachable workspace and the dexterous workspace [139]. The latter is the set that the end-effector frame can reach with arbitrary orientations, while the former is the set that the end-effector frame can reach with at least one orientation. Obviously, the dexterous workspace is a subspace of the reachable space. Both analytical and numerical methods have been used for charac-

G. Yang and I.-M. Chen, *Modular Robots: Theory and Practice*, Research on Intelligent Manufacturing,
https://doi.org/10.1007/978-981-16-5007-9_7

terizing these workspaces. Among the analytical approaches, a topological analysis of robot workspaces was given by Gupta and Roth [133], and Gupta [134]. In their works, the shapes of reachable workspace and dexterous workspace were analyzed. The concepts of workspace holes and voids were defined, and their existence conditions were identified. Further analytical studies of workspaces were addressed in Freudenstein and Primrose [135], where the relationships between robot kinematic parameters and workspace parameters were developed. This work was followed by an application to workspace optimization for 3-DOF robots [136]. A more general analysis of workspace optimization was presented by Paden and Sastry [131]. The optimal criteria of a serial type robot were defined in terms of the work-volume subject to a constraint on its length, and the well-connected workspace—the ability to reach all positions in its workspace in every pose. It was shown that the optimal design of a 6-R robot is an elbow manipulator, in which (1) the fourth, fifth, and sixth joint axes intersect at one point; (2) the Denavit-Hartenberg (D-H) twist angles are 0 or 90 degrees; and (3) either the link length offset or the joint offset is zero for each link. A geometric approach was given by Park [132], in which the workspace volume was precisely defined by regarding the rigid body motion as a Riemannian manifold. The significance of this work is that the so defined workspace volume depends only on the joint axes, while the lengths of links and the fixed reference frames need not be taken into account. Numerical studies of the robot workspace have been carried out by Kumar and Waldron [139], Yang and Lee [137], Tsai and Soni [138], and Yang and Hang [140]. These studies rely on numerical methods to generate projections of the workspace. The advantage of the numerical schemes over the analytical ones is that kinematic constraints can be easily included. However, the more general design principles and intuitions are more difficult to obtain, and the computation for the description of the workspace becomes very time consuming.

The manipulability or dexterity of a robot can be described as the ability to move and apply force and torque in arbitrary directions. One of the early papers to consider manipulability was the work of Salisbury and Craig [141]. The condition number of the Jacobian matrix (J), $\text{Cond}(J) = \sigma_{max}/\sigma_{min}$, was proposed to measure the transmission of the joint velocity errors to the fingertip velocity errors, where σ_{max} and σ_{min} are the maximum and minimum singular value of the Jacobian matrix, respectively. A clearer description of robot's manipulability was given by Yoshikawa [142], where the manipulability ellipsoid was introduced. The product of all singular values of the Jacobian matrix, which is equal to $\sqrt{\det JJ^T}$, was defined as the manipulability measure. This scalar quantity can measure how much the end-effector moves for a given infinitesimal movement of joint angles, averaged over all directions. In order to measure the closeness of a robot posture to a singularity, the Condition Index, $CI(J) = 1/\text{Cond}(J) = \sigma_{min}/\sigma_{max}$, has been proposed by Angeles [143]. Definitely, the range of CI is from 0 to 1. CI becomes zero if the robot is in a singular posture. A robot is called isotropic if its CI attains one. All the manipulability or dexterity measures mentioned above are local indices in the sense that they characterize the manipulability or dexterity of a robot at a given posture. They are useful for evaluating the motion ability of an existing robot. For the design of a general-purpose robot, however, global ones may be more desirable. A straightforward way that derives global dexterity measures is to integrate the local ones over some regions in the configuration space [144]. A geometric approach was given by Park [132], in which a global description of dexterity was obtained by introducing the harmonic mapping technique.

Besides the workspace volume and manipulability (or dexterity) measures, several other performance indices were also defined by different researchers such as the joint range availability [145], the solvability of the kinematic equations, the mechanical constructability [146], the accuracy achievability [147], and other nonlinearity and redundancy measures [148].

Most of the performance measures mentioned above are proposed for the analysis and design of general-purpose robots. For the design of a modular robot that is only aimed at performing a given task, the effectiveness of these performance measures remains unclear and needs to be studied. For example, the workspace volume may be a main performance index for a general-purpose robot. It becomes less important for evaluating a modular robot that is designed to cater for the specific task (if the task can be well executed).

Past researches on the optimal design of modular robots mainly focused on the so-called Task-Based Design (TBD) problem. The goal of TBD is to design a robot that is optimally suited to perform a given task. The studies on TBD were introduced by Paredis [32] for the Reconfigurable Modular Manipulator System (RMMS) [10]. In this work, the TBD problem was divided into several stages, i.e., kinematic design, dynamic design, controller design, and path planning. A robot task was defined as a set of working poses (positions and orientations) in the task space. The simulated annealing algorithm was employed to search for the optimal solution. The objective function was a penalty function in which penalties for violating task constraints, such as reachability and obstacle constraints, were combined in a weighted sum fashion.

A more comprehensive study on the design of task specific fault tolerant manipulator was also given by Paredis [22]. Such a TBD problem was also formulated as a design optimization problem. The task description consisted of a timed Cartesian path to be followed by the end-effector, and obstacle models in the manipulator workspace. An integrated evaluation function was proposed to deal with the tight coupling between the performance measures including joint angle, velocity, and torque limits, reachability, singularity avoidance, collisions with obstacles, and self-collision caused by the fault tolerance requirement. The objective criterion was defined as a desired trajectory with minimum energy consumption. Since many performance measures were considered, the evaluation procedure was very time consuming. To make this integrated approach computationally feasible, two techniques were addressed, i.e., to include problem specific design knowledge in the search, and implement the search in an easily parallelizable agent-based paradigm on the basis of Genetic Algorithms (GAs). Even so, the computational cost was still expensive. As mentioned in [36], to derive a feasible solution (with 75 percent probability) for the satellite docking task needs about 6 h of CPU time on twenty-four Sparc workstations.

Kim and Khosla [31, 149] also focused on the TBD problem for the RMMS. They posed the kinematic design of modular robots as an optimization problem. Similarly, the task requirement was also defined as a set of working poses in the task space. A Multi-Population Genetic Algorithm (MPGA) was employed as the optimization algorithm. The idea was to have one GA per task pose to search for locally optimal designs, and then progressively increased the coupling between the individual GAs to arrive at one single globally optimal design for all the task poses. The design variables included the number of DOFs, D-H parameters, and the base position. The dexterity measure was chosen as the objective criterion. Several kinematic constraints, such as dimension constraints, joint constraints, and task specification constraints, were considered.

Chen [14, 27] formulated the task-oriented optimal configuration problem as a combinatorial optimization problem due to the discrete feature of modules. The formulation was based on an Assembly Incidence Matrix (AIM) representation of a modular robot. The task requirement was defined as a set of sequential working points (positions) in the task space. A task dependent objective function, termed the Assembly Configuration Evaluation Function (ACEF), was proposed. This objective function contained both task evaluation and structure evaluation criteria. An Evolutionary Algorithm (EA) based on genetic operators defined on AIMs was employed to solve this optimization problem [27].

The task optimization problem was also addressed by Ambrose [1] for the modular robot system built in University of Texas at Austin. The purpose of this work was to allow a user to specify the robot performance requirements and then to select the best of the available module combinations for a set of goals. Ten criteria were considered in order to evaluate the performance of a robot, i.e., mobility, payload, speed, workspace, weight, backlash, friction, stiffness, inertia, and bandwidth. Several optimization algorithms such as the exhaustive search, the sequential filters, the modified golden section, the steepest descent, the conjugate direction, and Taguchi methods were employed for the design of one and two-dimensional robots for the comparison purpose, but no explicit objective function was defined.

Besides the task requirements, most of previous works on the TBD problem still emphasized robot performance measures such as manipulability, singularity avoidance, and fault tolerance, thus tending to select a sophisticated robot configuration with more DOFs as the optimal one. The reason behind this is that a robot with more DOFs can easily satisfy the task requirements and performance measures. An optimal robot under such considerations is usually a redundant one, e.g., more than 6 DOFs for a 3-dimensional manipulator and more than 3 DOFs for a regional (positioning) robot arm. For most industrial applications, however, the operation speed is the main concern. A modular robot with the simple kinematic structure—few modules and DOFs, is much more preferred in order to perform a given task swiftly. Therefore, how to design the Minimized Degrees of Freedom (MDOFs) modular robot for a given task is crucial to a modular robot system.

7.2 General Design Methodology

As mentioned in [22], the optimal design of a modular robot configuration for a given task is a very difficult problem. It involves both type (topological structure) synthesis and dimension synthesis. The design space is discrete and grows exponentially with the number of modules in the inventory. Various performance measures are highly coupled and nonlinear, and to evaluate whether the constraints are satisfied and to which extent the optimality criteria are achieved are very computationally expensive.

To reduce the complexity of TBD, we divide the design problem into several stages as shown in Fig. 7.1. The kinematic design is the first stage, i.e., Stage (1), because of its importance. It determines the kinematic configuration of the robot based on the kinematic requirements of a given task. The output of this stage is one "optimal" and several "sub-optimal" robot configurations. Since we aim at designing simple robots with MDOFs, it is more difficult to perform trajectory planning in the task space (Cartesian space). Instead, a joint space trajectory planning algorithm is to be employed for the derived robot configurations in Stage (2). The joint angle, velocity, and acceleration limitations thus can be easily considered in this stage. After that, a 3-D graphic simulation, based on the derived joint trajectory, will be given for the visualization of task execution and collision detection in Stage (3). Then, the computation of inverse dynamics will be performed for the investigation of joint torque limitations and the controller design in Stage (4). If the "optimal" robot configuration can successfully pass Stages (2), (3) and (4), it is definitely accepted as the final design; otherwise, the best one among the "sub-optimal" robot

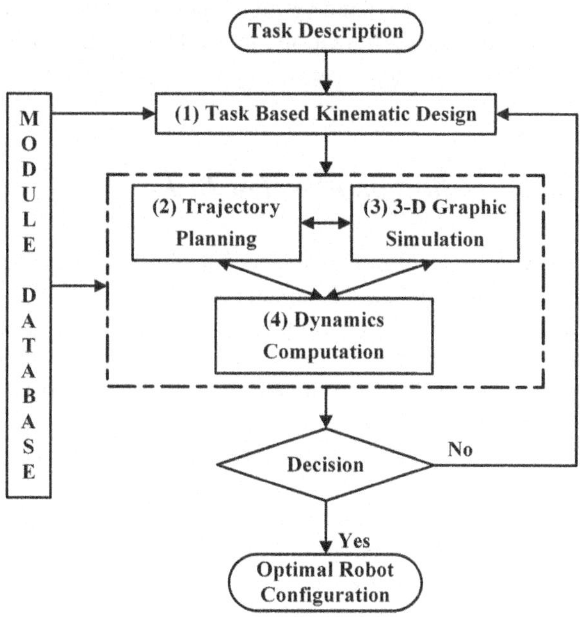

Fig. 7.1 Flow chart of the iterative design methodology

configurations will be selected and cast into the last three stages. This procedure will repeat until a survived one is obtained. If all the "sub-optimal" ones still fail to pass Stages (2), (3) and (4), the whole design procedure should be restarted and repeated until a desired robot configuration is derived. If it is the case that an acceptable solution cannot be found after a number of redesigns, it may imply that the given task is unreasonable and needs to be modified.

Note that the module specifications listed in Chap. 2 provide the kinematic and dynamic parameters of modules. For the purpose of collision detection, a 3-D graphic model is also to be created for each module.

To a certain extent, this design methodology is an iterative method. Iterations may occur to the whole design procedure or within several stages. It is true that such a method cannot guarantee that a desired robot configuration can be obtained within a few iterations. Compared to an integrated design method [22], however, this method can avoid spending unnecessary computation time to evaluate kinematically illegal individuals. This design methodology is based on the consideration that an optimal robot configuration should be kinematically optimal or sub-optimal one because the kinematic performance of a robot is critical for the task realization. Hence, in most cases, the proposed method is more efficient and practical. In the following sections of this chapter, we will mainly focus on the optimal kinematic design of modular robots for it is the most important part of the whole design procedure.

7.3 Optimization Model

The primary issue in the optimal kinematic design of a modular robot is to formulate a task-oriented optimization model. This model consists of three main parts: design parameters, the objective function, and constraints.

7.3.1 Definition of Robot Tasks

There are many different levels of details at which a robot task can be defined [22]. As an example, consider the assembly of a Printed Circuit Board (PCB). At the highest level, this task could simply be described as: "assemble a PCB". At an intermediate level, the task will be decomposed into many subtasks such as "locate a chip in the parts feeder", "pick up the chip", and then "insert the chip into the board", etc. At the lowest level, all the details are included, such as "move the end-effector from point A to point B", "exert a force of n Newtons on the end-effector in x direction to grasp a chip at point B", and "move the end-effector from point B to point C", etc. If the robot task is described at higher levels, a task planner is needed to convert the higher level task descriptions into the low level task descriptions. Although the task planning problem is also a challenging issue, it is not the main concern of this book.

Here, we would like to adopt a low level task description such that the robot motion sequence and trajectory are provided for task execution.

The execution levels of the robot motion sequence and trajectory are adopted to define a robot task for two reasons. First, with a prescribed robot motion sequence and trajectory, one can immediately determine the capability of a modular robot configuration to carry out the task. Furthermore, if the robot is unable to follow the motion, it is unnecessary to proceed any further with the evaluation procedure. Secondly, it is very difficult to define a measure to quantify the "goodness" of a robot while performing a task with a very vague definition [14].

For simplicity, a robot task is specifically defined as a collection of working points or poses, $\mathbf{w}_p$, to be followed by the end-effector or the end link in the task space [14], where $\mathbf{w}_p \in \Re^{3\times 1}$ or $\mathbf{w}_p \in SE(3)$. If the robot is to follow a prescribed path, this task can be approximated by a set of points or poses along the path. Other task specifications such as obstacle avoidance, position accuracy, and static capability at task points or poses can also be included as part of the definition of a task. Note that in the following context, the term "point" refers to the position of the end link frame and "pose" refers to both position and orientation of the end link frame.

7.3.2 Design Parameters and the Search Space

The parameters that determine a modular robot configuration are: (1) the number of the constituting link modules—N, (2) the link module types—L ($L \in \{L_{ci}, L_{ri}, L_{pi}\}$, $i = 1, 2$), and (3) the relative assembly orientations (port vectors) of the consecutive modules—P. When these parameters are all specified, a robot configuration is uniquely defined. Hence, we consider the set of design parameters X as:

$$X \to \{N, L, P\}. \tag{7.1}$$

To fully employ the power of the discrete optimization technique in the MDOF problem, we introduce a virtual link module type and a virtual joint type (connecting to the output socket of the virtual link module), denoted by L_υ and J_υ, respectively, into the inventory of modules. The virtual links and joints function as "slack parameters" to compensate for the fixed size representation of the candidate solutions for certain search/optimization techniques like GA. Both L_υ and J_υ do not have physical dimensions and do not exist in the actual robot assembly. We assume that the virtual link modules (L_υ) exist in the abstract kinematic description of the robot and have the virtual appearance of cubic link modules. To illustrate this virtual module concept, Fig. 7.2 shows the kinematic graph representation of three different modular robots. Physically they have different numbers of modules, i.e., five, four, and three modules, respectively, but the dimensions of their kinematic graphs as well as AIMs remain identical due to the introduction of the virtual modules.

According to the definition of the AIM, all the design parameters N, L, and P can be clearly indicated in an AIM. For every possible design (N,L,P), there exists

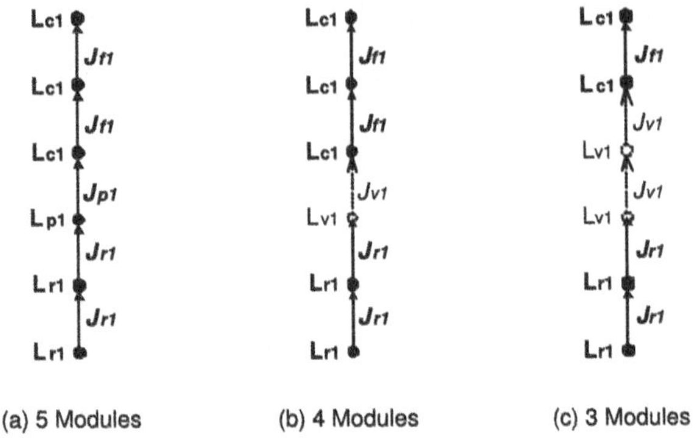

Fig. 7.2 Virtual module in the kinematic structure

a unique AIM. Therefore, the design parameters can be defined in AIM format. The size of the search space is equal to the number of all possible AIMs. Since the joints (connectors) depend on the links they are connected to, they are not independent parameters and are not considered as design parameters. Therefore, the last row of the AIM can be eliminated.

Excluding the virtual module, six kinds of modules ($L \in \{L_{ci}, L_{ri}, L_{pi}\}$, i=1, 2) can be chosen for robot assembly. Each module has six connecting sockets, and a module can be connected in four different orientations relative to its preceding module even if the connecting sockets remain unchanged. The search space thus becomes extremely large. For example, an assembly of two connected modules will have a total of $6 \times 6 \times 4 \times 6 \times 6 = 5184$ possible configurations. Likewise, consider an n-module serial type robot without any virtual module. The number of all possible configurations, N_{pc}, can be given by

$$N_{pc} = 4^{(n-1)} \times 5^{(n-2)} \times 6^{2n}. \tag{7.2}$$

Equation (7.2) shows that the search space grows exponentially with the number of modules in the robot. If the maximum allowed number of modules in a robot is equal to m and suppose that a robot consists of at least two real modules, such that the number of virtual modules may be 0, 1 . . . , or $m-2$, then the total number of possible configurations can be calculated by

$$N_{total} = \sum_{i=2}^{m} 4^{(i-1)} \times 5^{(i-2)} \times 6^{2i}. \tag{7.3}$$

Taking $m = 6$ as an example, the total number of the possible assembly configurations is 1.4×10^{15}. Note that in these configurations, there are some kinematically identical ones which are not eliminated from Eqs. (7.2) and (7.3). In [14], Chen proposed an algorithm that can enumerate all of the distinct modular robot configurations from a given set of modules. Chen's enumeration algorithm can effectively reduce the size of the search space so that it is more suitable for the case of exhaustive search.

Nevertheless, obtaining an optimal robot configuration from such a large and discrete search space is definitely time consuming.

According to the module designs, when we assemble a complete robot with different sizes of modules, a general rule is that the large modules should precede the small modules which are usually located at the distal ends. In the search space, however, there still exist such robot configurations that: (1) small size modules are connected between two large size modules; (2) large size modules are connected between two small size modules. From the mechanical point of view, a robot with such assembly patterns is an unreasonable structure because of the higher energy consumption and lower loading capability. To avoid this problem, we propose here a two-step configuration design method: the optimization step and the module replacement step.

In the first step, only the large size modules are considered in the inventory of modules, i.e., $L \in \{L_{c1}, L_{r1}, L_{p1}, L_v\}$, so that many mechanically infeasible configurations can be avoided and the search space will be significantly reduced. The total number of the possible assembly configurations can be computed by

$$N_{total} = \sum_{i=2}^{m} 4^{(i-1)} \times 5^{(i-2)} \times 6^i \times 3^i. \tag{7.4}$$

Also taking $m = 6$ as an example, the possible assembly configurations are reduced to 2.2×10^{13} (about one-sixtieth of the size of the original search space).

In the second step, the large modules in the optimal (or sub-optimal) robot configuration will be replaced with small modules of the same types as many as possible. The module replacement will be done from the end link module to the base link module one by one. Each module replacement will also satisfy the computation results of Stages (2), (3), and (4) in Fig. 7.1 and the assembly rules to be mentioned in Sect. 7.4. Moreover, when we replace a large module with a small one in a robot, the length of the corresponding connectors should be enlarged accordingly in order to remain the kinematic parameters unchanged. It follows that after the module replacement step, the kinematic structure of the robot still remains the same. The resulting robot configuration, on the other hand, becomes more reasonable for the given task. Since this chapter focuses on the kinematic design of modular robot configurations, only the first step, the optimization step, is addressed here.

7.3.3 Objective Function

The objective function is to evaluate the "goodness" of a robot assembly configuration for a given task. Since the performance measures are defined to quantify the features of robot performance, a straightforward way to define the objective function is to list all necessary performance measures such as reachability, manipulability, and fault tolerance index, and then choose an important one among them as a single objective function [14, 27, 31] or combine some or all of them in a weighted sum as a multi-objective function [29, 32]. Performance measures that are not included in the objective function can be treated as constraints. However, if the objective function only focuses on robot performance measures, there will be a tendency of selecting a robot configuration with more DOFs as the optimal one. Moreover, since the type (topology) synthesis will also be performed in the optimization procedure, some commonly used performance measures become invalid when used as objective criteria. For example, the manipulability and workspace measures are both powerful criteria for evaluating robots with the same topological structure, but it is unfair when using them to evaluate robots with different topological structures.

Because of the modular design, a modular robot will have most of its actuators being located on the moving arm itself. Using excessive modules, especially the actuators, will certainly lower the robot's loading capability. Hence, to use a minimal number of modules as well as DOFs to construct a robot assembly for a task becomes our main concern. The objective function F is therefore defined as a weighted sum of the numbers of different types of modules to be employed in a robot:

$$F = k_{p1} N_{p1} + k_{r1} N_{r1} + k_{c1} N_{c1}, \tag{7.5}$$

where N_{p1}, N_{r1}, and N_{c1} represent the numbers of large size prismatic, revolute, and cubic link modules, respectively; k_{p1}, k_{r1}, and k_{c1} are the user-defined weightages of N_{p1}, N_{r1}, and N_{c1}, respectively. All these weightages are positive constants and reflect the relative contribution of their corresponding link modules to the objective function. To design MDOF robots, the value of the objective function is to be minimized. In this case, the weightages in Eq. (7.5) would actually act as penalty coefficients. When we wish to minimize the number of revolute and prismatic joints (DOFs), we will set k_{r1} and k_{p1} to be much larger than k_{c1}, as cubic link modules do not contribute to the DOFs of the robot. If we wish to have more revolute joints than prismatic joints (i.e., $N_{r1} > N_{p1}$), then the weightage k_{p1} is set to be greater than k_{r1}; if we treat both types of joints to be of equal importance, then k_{p1} is set to be equal to k_{r1}; otherwise, k_{r1} is set to be greater than k_{p1}. These basic rules are only for determining the relative importance of the types of modules. Suitable values of k_{p1}, k_{r1}, and k_{c1} have to be determined by trial computations.

Since a classical GA maximizes the objective function during the search procedure, we define the reciprocal of $F (F > 0)$, $1/F$, as the fitness function for the search algorithm. Without loss of physical interpretation, the term "fitness" in the context still refers to the objective function value as given by Eq. (7.5).

7.3.4 Performance Constraints

The performance constraints are defined to ensure the feasibility of a robot configuration while performing the given task. The constraints considered here are:

- Reachability;

- Joint Range Availability;
- Manipulability;
- Mechanical Constructability.

The **reachability constraint** is used to inspect whether all of the given task points/poses are within a robot's reachable workspace. If this constraint is satisfied, the robot configuration is able to perform the given task kinematically. To study the reachability constraint of a robot, we transform the problem into the inverse kinematics problem. Because a modular robot has unfixed assembly configurations, the numerical inverse kinematic algorithm presented in Chap. 4 is employed. For a candidate robot, if an inverse solution is found at each of the task points/poses, the reachability constraint is considered to be satisfied. Suppose that each of the given task points or poses is $w_{pi} \in \Re^{3\times 1}$ or $SE(3)$ $(i = 1, 2, \ldots, m)$, the reachability constraint for a robot with $n + 1$ modules is given by:

$$T_{0,1}(0)e^{\hat{s}_1 q_{1i}} T_{1,2}(0)e^{\hat{s}_2 q_{2i}} \ldots T_{n-1,n}(0)e^{\hat{s}_n q_{ni}} = w_{pi}. \tag{7.6}$$

The **joint range availability constraint** is defined to detect whether the joint angles derived at each task point/pose are within the joint angle limitations. The joint range availability constraint for a robot with $n + 1$ modules is then given by

$$a_i \geq q_i \geq b_i (i = 1, 2, \ldots, n), \tag{7.7}$$

where b_i and a_i are the lower bound and upper bound of the joint i's displacement, respectively.

The **manipulability constraint** here is only used to detect whether a robot is at or nearby a singular posture when its end-effector reaches each of the task points or poses, and thus it is a local manipulability index. For convenience, the CI is employed to formulate the manipulability constraint:

$$\sigma_{min}/\sigma_{max} \geq \epsilon, \tag{7.8}$$

where σ_{min} and σ_{max} are the minimum and maximum singular values of the robot Jacobian matrix, respectively. ϵ is the user defined lower bound of CI for singularity avoidance. It usually takes a small value, 0.001 for instance.

The **mechanical constructability constraint** is proposed to check the mechanical feasibility of a robot assembly. For example, structures with two or more modules attached to the same connecting socket of an identical module are obviously infeasible. To avoid such structures, mechanical constructability constraints have to be given. However, if there are excessive constraints, one will run the risk of creating a search algorithm that spends most of its time evaluating illegal individuals [150]. Instead of formulating these constraints, several assembly rules are built into the search algorithm so that mechanically infeasible structures can be intelligently avoided. The assembly rules for fixed base, serial type robots are:

1. A complete modular robot consists of at least two real modules.

2. The local coordinate frame of the base module coincides with the world frame.
3. For each of the intermediate modules, only two out of the six connecting sockets will actually be used for assembly, i.e., the "input socket" which is connected with its proceeding module and the "output socket" which is connected with its succeeding module.
4. The base module only has an "output socket", and does not have any "input socket". On the contrary, the end module has an "input socket", and does not have any "output socket".
5. The moving socket of any joint module is always used as "output socket" such that the z direction of the module frame points to its succeeding module.

The first four assembly rules focus on constructing a robot with a fixed base, serial type kinematic structure. The last rule is aimed at assembling a robot with a reasonable mechanical structure. Taking the fifth rule as an example, if the "output socket" of a revolute joint module is not the moving socket, it is functionally the same as a link module, but it is heavier and more expensive than the link module and is thus definitely unacceptable. Since the coding schemes for search algorithm are defined on AIMs, these assembly rules will inherently act on AIMs through an AIM generating scheme and genetic operators (mutation operators). As a result, the search algorithm will always evaluate mechanically feasible robot configurations.

7.4 Evolutionary Algorithm

Since both the design parameters and the search space are discrete in nature, combinatorial optimization techniques can be applied. Exhaustive search techniques can find the exact optimal solution, but the search space is so large that implementation of such an algorithm is inefficient. Random search techniques such as the GA and the Simulated Annealing (SA) method are more suitable for such a problem [14, 35].

The EA is a variation of GA with problem specific data representation and genetic operators [150]. It is a probabilistic search method based on the principle of evolution and hereditary of nature systems. In such an algorithm, a population of individuals for each generation is maintained. The individual is implemented with some data structure and is evaluated by a "fitness function" to give a measure of its "fitness". A new population is formed by selecting the more suitable individuals. In this selection procedure, an individual with larger "fitness" value ($1/F$) is more likely to be selected for the new generation. Some members of the new generation undergo transformations by the "genetic operators" to form new solutions. Probability rules are applied to determine the execution of the operators. Two types of genetic operators are considered in general: crossover type and mutation type. The crossover operation reorganizes data segments in several individuals to form new individuals (or the offspring). The mutation operator makes small changes in a single individual. After some number of generations, the individuals will converge to the optimal or nearly optimal solution.

7.4.1 Coding Scheme

It is reported that the computational efficiency and stability of GAs rely primarily on the data representation method, i.e., the coding scheme [150]. Classical GAs usually use fixed-length binary strings as a chromosome (the data structure) for its individuals and the genetic operators: the binary mutation and the binary crossover. Such a coding scheme when adopted for robot configuration designs becomes inconvenient for the binary string representation is not sufficient to reflect the nature of the design parameters. Also, a conversion scheme is required to transform the binary strings into the AIMs [14]. In an EA, on the other hand, chromosomes need not be represented by binary strings. The alteration process includes employing the problem-specific data structure to represent the chromosomes and modifying the "genetic" operators appropriately for the given data structure. Since the AIM representation conceptually defines the data structure of the solution space, we adopt the AIM as the chromosome and define the AIM related genetic operators. The main objective behind such an implementation is to move the EA closer to the solution space.

7.4.2 AIM Generating Scheme

The AIM generating scheme is proposed to automatically generate the initial population of AIMs. This scheme consists of two parts: link type generation and port vector generation. Although both link types and port vectors are generated on the basis of the assembly rules, the generation procedure would still have a random feature. For example, based on the first and second assembly rules, we can still randomly choose either L_{c1}, L_{r1}, L_{p1}, or L_v as the base link. The input of this scheme is an inventory of link modules—$L = \{L_{c1}, L_{r1}, L_{p1}, L_v\}$ and the maximum allowable number of modules to be used in a robot—n_m. The output is an AIM satisfying all the assembly rules.

7.4.3 Genetic Operators on AIMs

7.4.3.1 Crossover Operator

The crossover operator applies to the rows of two AIMs. A cut-off row is randomly chosen. Then the rows in the two AIMs below the cut-off row (Fig. 7.3) are swapped to form two new AIMs. The probability p_c controls this operation.

Before Swaping

$$A1 = \left[\begin{array}{cccc|c} (z,x) & 0 & 0 & 0 & Lp1 \\ (x,y) & (z,y) & 0 & 0 & Lc1 \\ 0 & (x,y) & (z,-y) & 0 & Lp1 \\ \hline 0 & 0 & (-z,-y) & (z,-y) & Lr1 \\ 0 & 0 & 0 & (-z,x) & Lc1 \end{array}\right] \qquad A2 = \left[\begin{array}{cccc|c} (z,x) & 0 & 0 & 0 & Lr1 \\ (-x,z) & (z,-y) & 0 & 0 & Lr1 \\ 0 & (-x,-y) & (z,x) & 0 & Lc1 \\ \hline 0 & 0 & (y,z) & (z,x) & Lp1 \\ 0 & 0 & 0 & (y,x) & Lc1 \end{array}\right]$$

After Swaping

$$A1' = \left[\begin{array}{cccc|c} (z,x) & 0 & 0 & 0 & Lp1 \\ (x,y) & (z,y) & 0 & 0 & Lc1 \\ 0 & (x,y) & (z,-y) & 0 & Lp1 \\ \hline 0 & 0 & (y,z) & (z,x) & Lp1 \\ 0 & 0 & 0 & (y,x) & Lc1 \end{array}\right] \qquad A2' = \left[\begin{array}{cccc|c} (z,x) & 0 & 0 & 0 & Lr1 \\ (-x,z) & (z,-y) & 0 & 0 & Lr1 \\ 0 & (-x,-y) & (z,x) & 0 & Lc1 \\ \hline 0 & 0 & (-z,-y) & (z,-y) & Lr1 \\ 0 & 0 & 0 & (-z,y) & Lc1 \end{array}\right]$$

Fig. 7.3 Crossover operation

$$\begin{bmatrix} (z,x) & 0 & 0 & 0 & Lr1 \\ (-x,z) & (z,-y) & 0 & 0 & Lr1 \\ 0 & (-x,-y) & (z,x) & 0 & Lc1 \\ 0 & 0 & (-z,-y) & (z,-y) & \boxed{Lr1} \\ 0 & 0 & 0 & (-z,y) & Lc1 \end{bmatrix} \xrightarrow{\text{Link Type Mutation}} \begin{bmatrix} (z,x) & 0 & 0 & 0 & Lr1 \\ (-x,z) & (z,-y) & 0 & 0 & Lr1 \\ 0 & (-x,-y) & (z,x) & 0 & Lc1 \\ 0 & 0 & (-z,-y) & (z,-y) & \boxed{Lc1} \\ 0 & 0 & 0 & (-z,y) & Lc1 \end{bmatrix}$$

$$\xrightarrow{\text{Port Vector Mutation}} \begin{bmatrix} (z,x) & 0 & 0 & 0 & Lr1 \\ (-x,z) & (z,-y) & 0 & 0 & Lr1 \\ 0 & (-x,-y) & (z,x) & 0 & Lc1 \\ 0 & 0 & (-z,-y) & (z,-y) & Lr1 \\ 0 & 0 & 0 & (-z,x) & Lc1 \end{bmatrix}$$

Fig. 7.4 Mutation operation

7.4.3.2 Mutation Operators

The mutation operators act on non-zero entries of a single AIM. Since the non-zero entries in a modified AIM (without the last row) can be divided into two categories: port vectors and link types (the last column), two types of mutations for both port vectors and link types are considered.

- **Link type mutation:** As shown in Fig. 7.4, the link type mutation acts on the last column of an AIM. The link type will be altered randomly, while the port vectors remain unchanged. The probability p_{ml} controls this operation. This operation will be subject to the assembly rules.
- **Port vector mutation:** As shown in Fig. 7.4, port vector mutation acts on the port vectors of an AIM. The port vector is altered randomly, while types of links remain unchanged. The probability p_{mp} controls this operation. This operation will also be subject to the assembly rules.

Fig. 7.5 The kinematic design algorithm

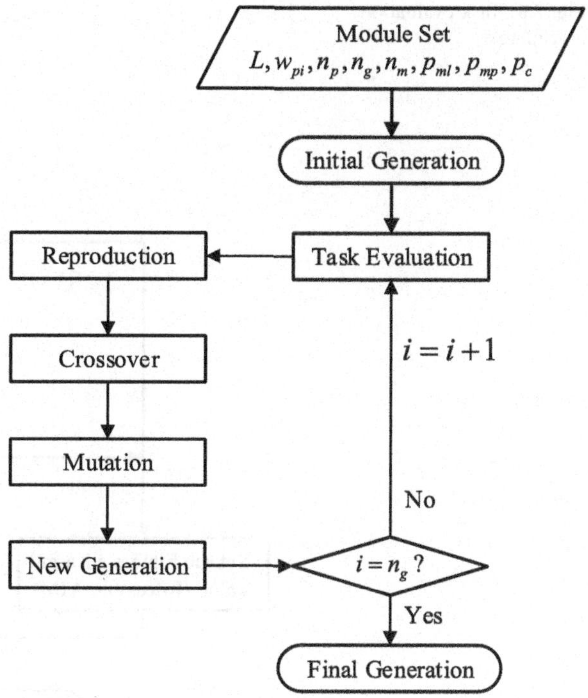

7.4.4 Implementation of the Evolutionary Algorithm

The proposed EA is successfully implemented in *Mathematica*. As shown in Fig. 7.5, the input parameters include an inventory of link modules L, the population size n_p, the number of destination generation n_g, the maximum allowable number of modules n_m, the crossover operator p_c, and the mutation operators p_{ml} and p_{mp}. The first step is to randomly generate n_p AIMs for the initial generation by using the AIM generating scheme. It is followed by the task evolution procedure to compute the fitness value for each AIM (Fig. 7.6). After the task evolution procedure, a new generation of AIMs is produced through reproduction, crossover, and mutation operations. The entire process will be repeated until the predetermined number of generations n_g is reached. In the final generation, we choose the AIM with the smallest fitness value (F) as the optimal assembly configuration. Moreover, during the search procedure, the optimal one in each generation is also recorded. Several sub-optimal AIMs can thus be derived from the intermediate optimal configurations.

The task evaluation procedure, as shown in Fig. 7.6, consists of two parts: performance constraint detection and fitness value computation. If an AIM can pass all the constraints, i.e., reachability, joint range availability, and manipulability, its fitness can be calculated by Eq. (7.5); otherwise, the fitness is assigned to an infinitely large number to indicate that such an AIM is infeasible. In the real implementation, however, an infinitely large fitness value is not convenient for the reproduction

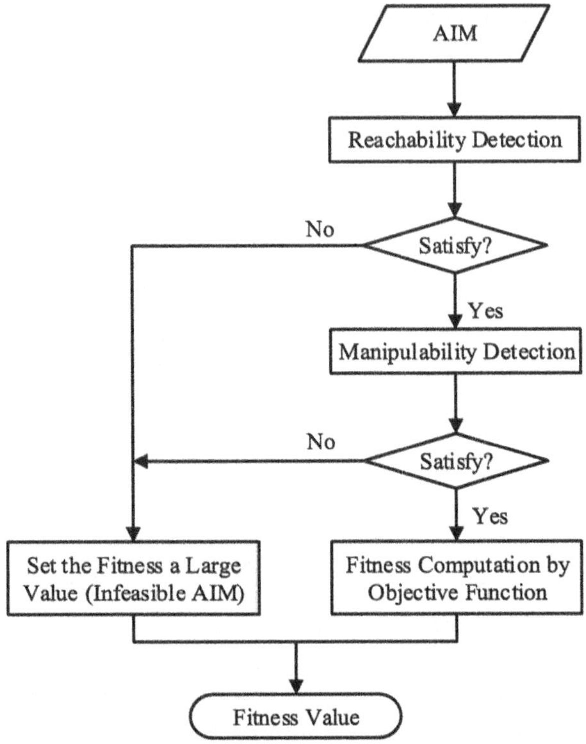

Fig. 7.6 Task evaluation procedure

(selection) procedure, so we replace it with a very large number to indicate that such an AIM consists of excessive modules and has little chance to be selected for the next generation.

Since the reachability constraint is very strict and is difficult to be satisfied, the EA may become unstable. It may be the case that the final generation does not contain any acceptable solution even if the optimal individuals have ever been created in the intermediate generations. Hence, two techniques are employed to increase the stability of the algorithm.

- **Adaptive mutation operators:** the mutation operators (p_{mp} and p_{ml}) are dynamically changed according to the average fitness of each generation. The suitable relationship between mutation operators and the average fitness values is determined through trial computations. In general, the smaller the average fitness value, the smaller the mutation operators. If all the individuals in a generation are infeasible configurations, then the probability of mutation operations becomes very high. Consequently, this is equivalent to randomly selecting AIMs.
- **Eugenics:** the best AIM obtained previously will be retained from generation to generation until a better one is created. Then the new AIM will replace the old one.

By employing these two techniques, the algorithm becomes more stable and can always find acceptable solutions. However, if the given mutation operators and the population size are too small, one will run a risk that the algorithm may converge to a local minimum.

7.5 Computation Examples

In order to demonstrate the effectiveness of the proposed EA, two examples are given in this section. In the first example, the given task is a set of working points (positions), termed the position design problem, while in the second example the given task is a set of working poses (positions/orientations), termed the pose design problem. In both examples, the user defined weightages for different types of link modules are taken as $k_{p1} = 1.5$, $k_{r1} = 1.0$, and $k_{l1} = 0.25$; the crossover operator is uniformly taken as $p_c = 0.25$. If an AIM cannot satisfy the constraints, its fitness is set to 20.

Example 7.1 (*Position Design Problem*) In this example, we wish to design a fixed base serial robot that can pass through a set of task points listed in Table 7.1. We set $n_m = 6$ and $n_g = 50$. The adaptive probabilities for the mutation operators are listed in Table 7.2.

The computation results are shown in Figs. 7.7, 7.8, 7.9, and 7.10. Figure 7.7 shows the initial generation of the robot configurations which consists of 20 different robot configurations. As shown in Figs. 7.9 and 7.10, both the average and minimal fitness values in the initial generation are equal to 20.0 which indicates that none of the initial generation of robot configurations is feasible. From the second generation, both average and minimal fitness values begin to decrease indicating that at least one feasible configuration has been found. When the destination generation, the 50th generation, is arrived, the average and minimum fitness values become 3.61 and 2.75, respectively. The final generation of robot configurations is shown in Fig. 7.8.

Table 7.1 Task points

Point 1	Point 2	Point 3	Point 4
(−212.52, −533.07, 0)	(247.48, −597.48, 0)	(641.18, −265.59,0)	(700.00,0,0)

Table 7.2 The adaptive probabilities for the mutation operators (position problem)

Average fitness value F_a	Link mutation operator	Port mutation operator
20	0.2	0.2
$10 \leq F_a < 20$	0.05	0.05
$5 \leq F_a < 10$	0.03	0.03
$3 \leq F_a < 5$	0.02	0.02
$F_a < 3$	0.015	0.015

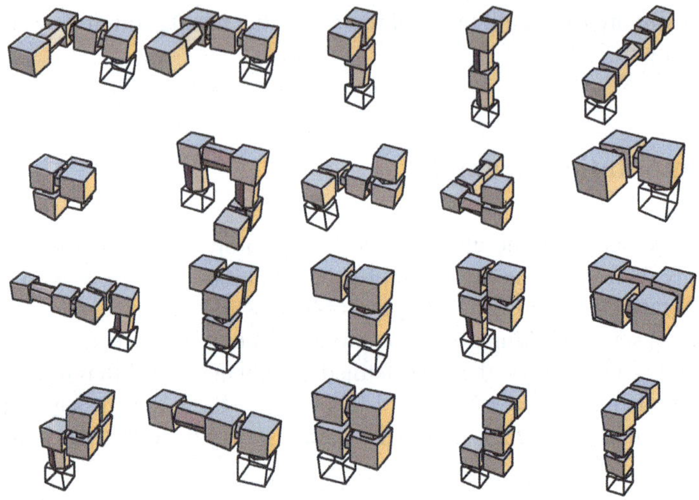

Fig. 7.7 Robot configurations in the initial generation (position problem)

All these configurations, except for the 12th configuration (which is an infeasible one), have the same kinematic structure. Each of them contains two revolute joint modules (2 DOFs), three cubic link modules, and one virtual module, and therefore has the fitness of 2.75 (the smallest fitness in all the 50 generations). Note that due to the simple schematic drawing routine, the cubic link module L_{c1} and revolute link module L_{r1} look alike. However, we can identify them through the resulting AIMs. The AIM of the optimal robot configuration is given by (Fig. 7.11)

$$\mathcal{A}_{\textbf{opt}} = \begin{bmatrix} (z,-x) & 0 & 0 & 0 & 0 & L_{r1} \\ (-x,-z) & (z,-y) & 0 & 0 & 0 & L_{c1} \\ 0 & (-x,y) & (z,x) & 0 & 0 & L_v \\ 0 & 0 & (-x,-y) & (z,-x) & 0 & L_{r1} \\ 0 & 0 & 0 & (-x,-z) & (z,-y) & L_{c1} \\ 0 & 0 & 0 & 0 & (-y,x) & L_{c1} \\ J_{r1} & J_{f1} & J_v & J_{r1} & J_{f1} & 0 \end{bmatrix}.$$

The execution of the given task (four points) with the optimal robot configuration is shown in Fig. 7.12. Three sub-optimal assembly configurations are also obtained as shown in Fig. 7.13. They have the fitness values of 5.25 (5 DOFs), 4.25 (4 DOFs), and 3.5 (3 DOFs), respectively. Their corresponding AIMs are given by

$$\mathcal{A}_{\textbf{sub1}} = \begin{bmatrix} (z,-y) & 0 & 0 & 0 & 0 & L_{r1} \\ (x,-z) & (z,y) & 0 & 0 & 0 & L_{r1} \\ 0 & (-x,y) & (z,-x) & 0 & 0 & L_{r1} \\ 0 & 0 & (x,-z) & (z,y) & 0 & L_{r1} \\ 0 & 0 & 0 & (y,z) & (z,x) & L_{r1} \\ 0 & 0 & 0 & 0 & (y,-x) & L_{c1} \\ J_{r1} & J_{r1} & J_{r1} & J_{r1} & J_{r1} & 0 \end{bmatrix};$$

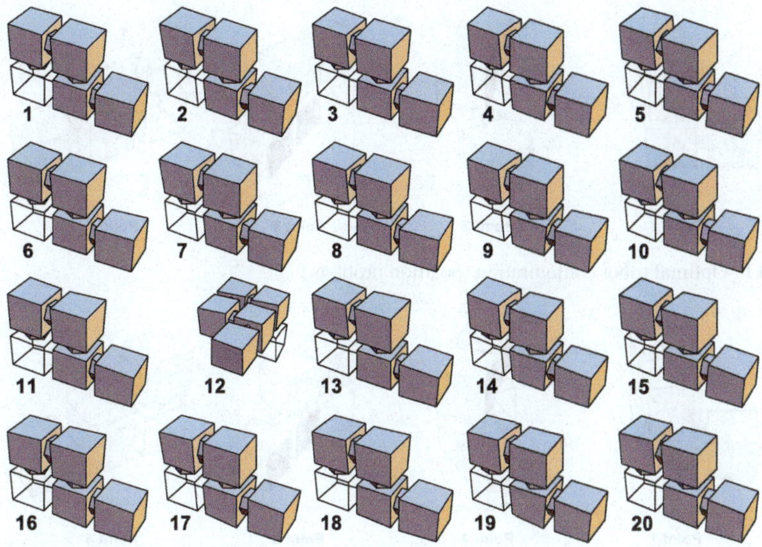

Fig. 7.8 Robot configurations in the final generation (position problem)

Fig. 7.9 Average fitness value in every generation (position problem)

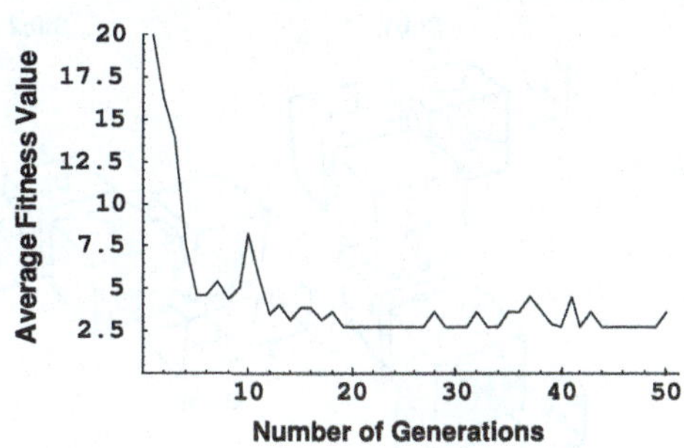

Fig. 7.10 Minimal fitness value in every generation (position problem)

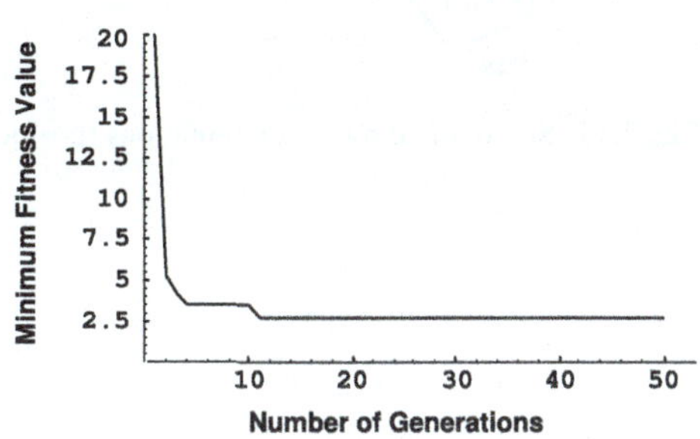

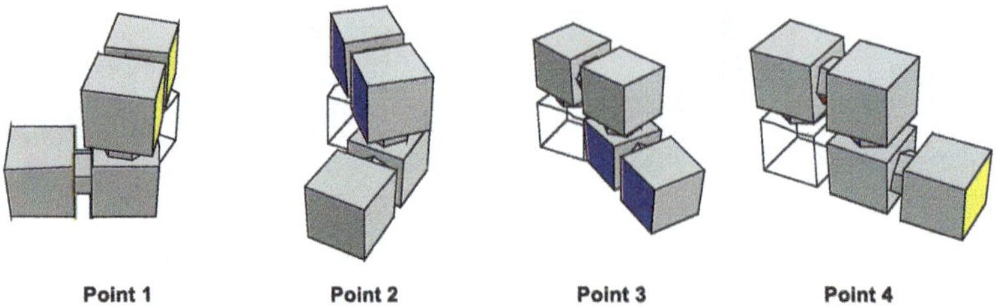

Fig. 7.11 Optimal robot configuration (position problem)

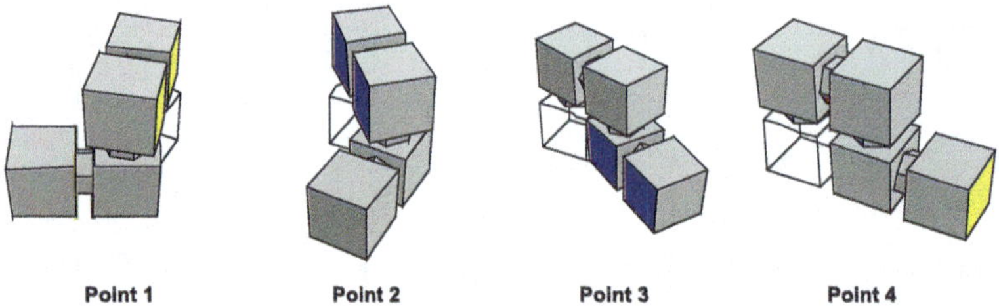

Fig. 7.12 Execution of the given task with the optimal robot configuration (position problem)

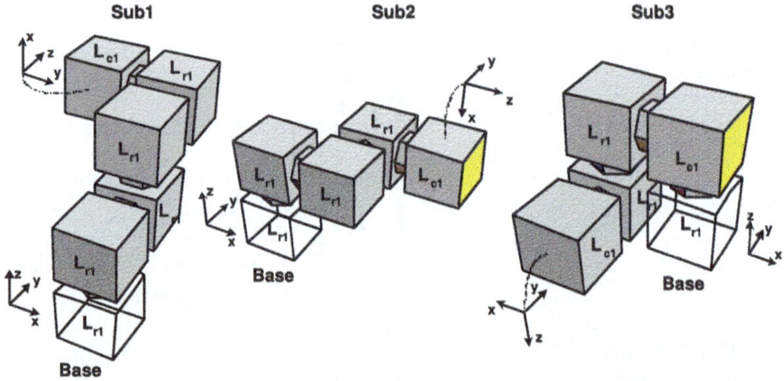

Fig. 7.13 Sub-optimal robot configurations (position problem)

$$\mathcal{A}_{\text{sub2}} = \begin{bmatrix} (z,-y) & 0 & 0 & 0 & 0 & L_{r1} \\ (-x,y) & (z,-x) & 0 & 0 & 0 & L_{r1} \\ 0 & (-x,y) & (z,-x) & 0 & 0 & L_{v} \\ 0 & 0 & (x,-y) & (z,x) & 0 & L_{r1} \\ 0 & 0 & 0 & (y,-z) & (z,x) & L_{r1} \\ 0 & 0 & 0 & 0 & (-z,-x) & L_{c1} \\ J_{r1} & J_{r1} & J_{v} & J_{r1} & J_{r1} & 0 \end{bmatrix};$$

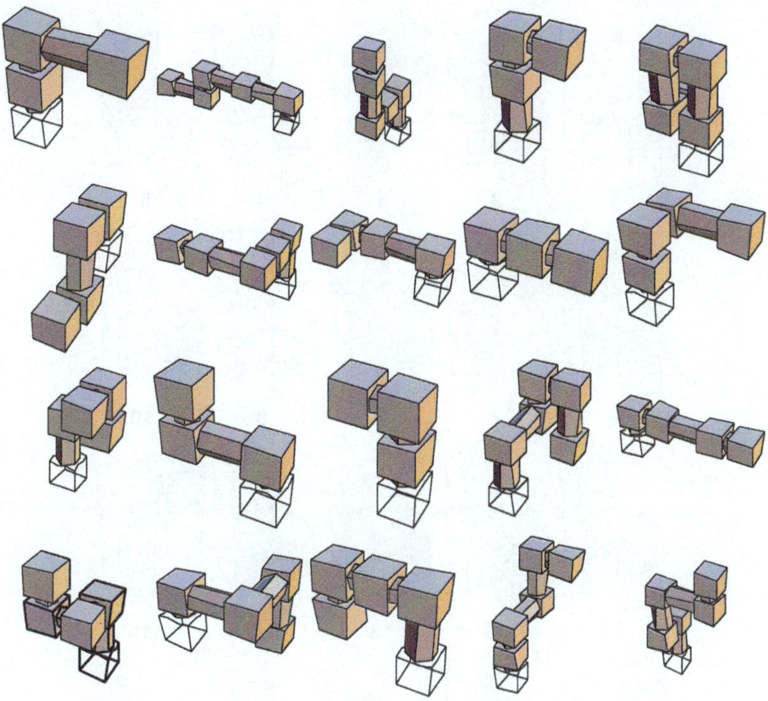

Fig. 7.14 Robot configurations in the initial generation (pose problem)

$$\mathcal{A}_{\mathbf{sub3}} = \begin{bmatrix} (z,-y) & 0 & 0 & 0 & 0 & L_{r1} \\ (-x,-y) & (z,x) & 0 & 0 & 0 & L_{c1} \\ 0 & (-x,y) & (z,x) & 0 & 0 & L_{v} \\ 0 & 0 & (-x,-z) & (z,-y) & 0 & L_{r1} \\ 0 & 0 & 0 & (y,z) & (z,x) & L_{r1} \\ 0 & 0 & 0 & 0 & (y,-x) & L_{c1} \\ J_{r1} & J_{f1} & J_{v} & J_{r1} & J_{r1} & 0 \end{bmatrix}.$$

Example 7.2 (*Pose Design Problem*) Now we wish to design the MDOF robot that can pass a set of task poses listed in Table 7.3. In this example, we set $n_m = 6$ and $n_g = 100$. The mutation operators for the search algorithm are listed in Table 7.4.

The computation results are shown in Figs. 7.14, 7.15, 7.16, and 7.17. Figure 7.14 shows the initial generation of robot configurations which contains 20 different robot configurations, and none of them is feasible because both minimal and average fitness values are equal to 20 (Figs. 7.16 and 7.17). From the 22nd generation as shown in Figs. 7.16 and 7.17, both the average and minimal fitness values begin to decrease, which indicates that at least one feasible configuration has been found. When the destination generation, the 100th generation, is reached, the average and minimal fitness values become 4.63 and 2.75, respectively. The robot configurations in the final generation are shown in Fig. 7.15. Both the 13th and 20th robot configurations in

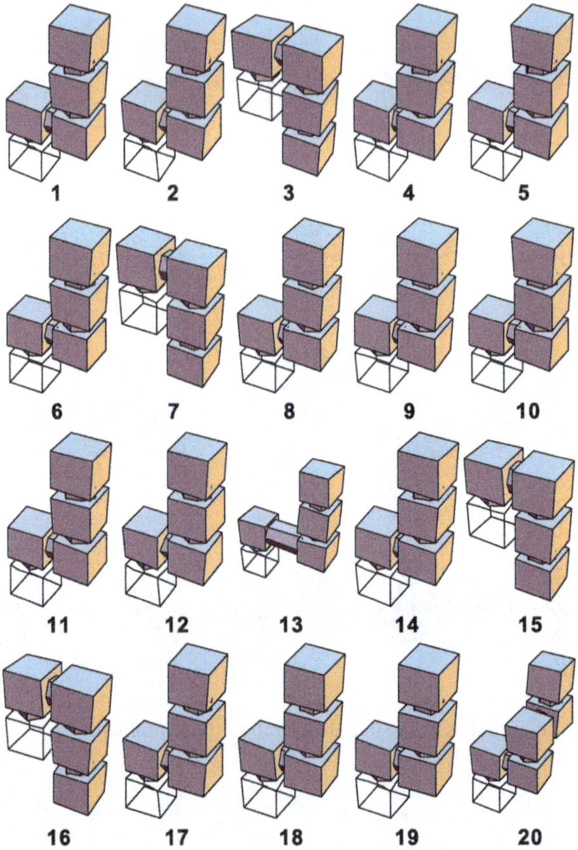

Fig. 7.15 Robot configurations in the final generation (pose problem)

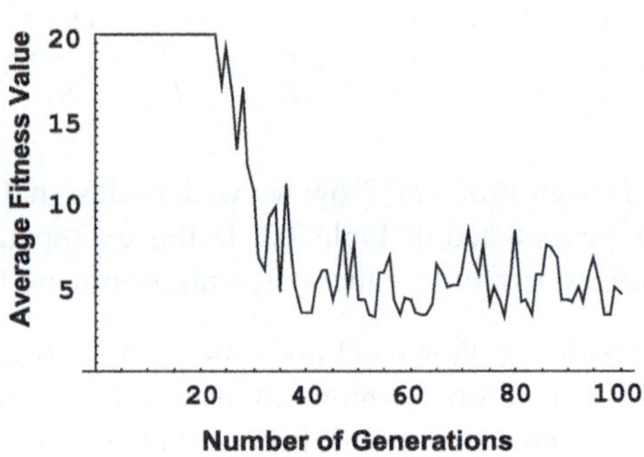

Fig. 7.16 Average fitness value in every generation (pose problem)

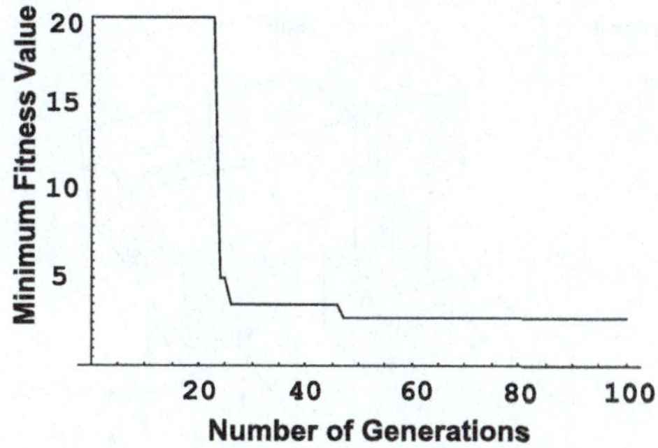

Fig. 7.17 Minimal fitness value in every generation (pose problem)

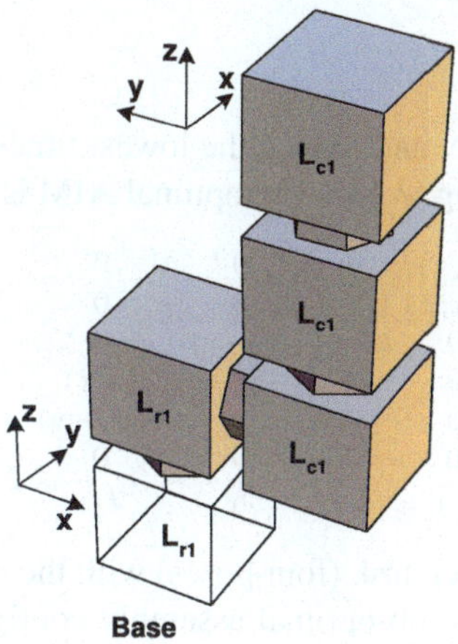

Fig. 7.18 Optimal robot configuration (pose problem)

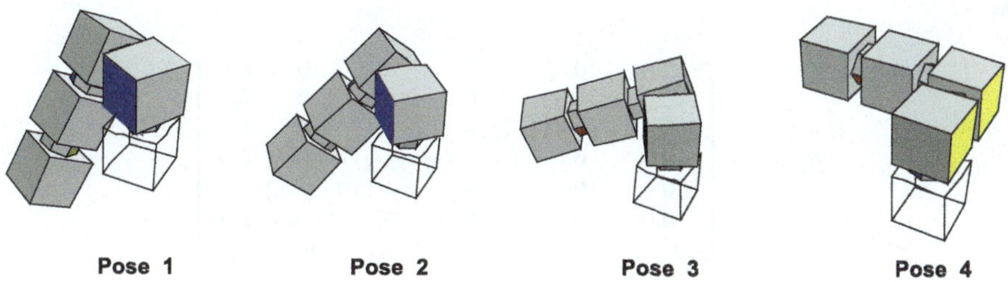

Fig. 7.19 Execution of the given task with the optimal robot configuration (pose problem)

Fig. 7.15 have the fitness value of 20 and are thus infeasible ones. The 3rd, 7th, 15th, and 16th robot configurations have the same kinematic structure and bear the fitness value of 3.5. From the AIM, we can see that this configuration contains three revolute link modules (3 DOFs), two cubic link modules, and one virtual link module. The reminders belong to another set of kinematically identical configurations with each of them consisting of two revolute link modules (2 DOFs), three cubic link modules,

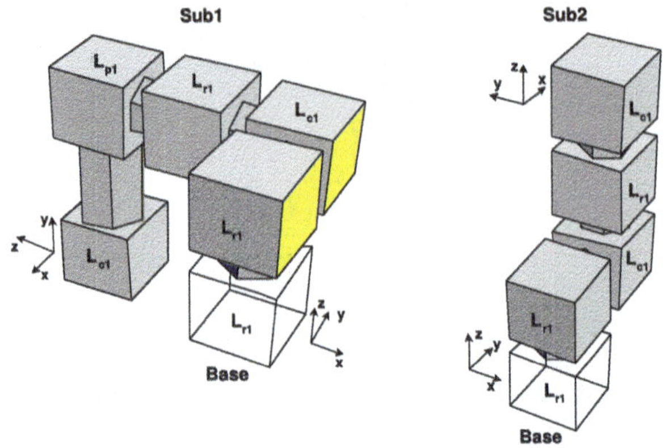

Fig. 7.20 Sub-optimal robot configurations (pose problem)

and one virtual link module, and having the lowest fitness value of 2.75, and hence become the optimal one (Fig. 7.18). The optimal AIM is

$$\mathcal{A}_{\mathbf{opt}} = \begin{bmatrix} (z,x) & 0 & 0 & 0 & 0 & L_{r1} \\ (-x,z) & (z,y) & 0 & 0 & 0 & L_{r1} \\ 0 & (x,y) & (z,-y) & 0 & 0 & L_{v} \\ 0 & 0 & (-z,-y) & (z,x) & 0 & L_{c1} \\ 0 & 0 & 0 & (-z,-y) & (z,x) & L_{c1} \\ 0 & 0 & 0 & 0 & (-z,x) & L_{c1} \\ J_{r1} & J_{r1} & J_{v} & J_{f1} & J_{f1} & 0 \end{bmatrix}.$$

The execution of the given task (four poses) with the optimal robot configuration is shown in Fig. 7.19. Two sub-optimal assembly configurations are also obtained from the intermediate generations as shown in Fig. 7.20, which have the fitness values of 5.0 (4 DOFs) and 3.5 (3 DOFs), respectively. Their corresponding AIMs are given by

$$\mathcal{A}_{\mathbf{sub1}} = \begin{bmatrix} (z,y) & 0 & 0 & 0 & 0 & L_{r1} \\ (y,z) & (z,-y) & 0 & 0 & 0 & L_{r1} \\ 0 & (-x,-y) & (z,-y) & 0 & 0 & L_{c1} \\ 0 & 0 & (-z,-y) & (z,y) & 0 & L_{r1} \\ 0 & 0 & 0 & (y,z) & (z,-x) & L_{p1} \\ 0 & 0 & 0 & 0 & (y,x) & L_{c1} \\ J_{r1} & J_{r1} & J_{f1} & J_{r1} & J_{p1} & 0 \end{bmatrix};$$

$$\mathcal{A}_{\mathbf{sub2}} = \begin{bmatrix} (z,y) & 0 & 0 & 0 & 0 & L_{r1} \\ (-y,z) & (z,-y) & 0 & 0 & 0 & L_{r1} \\ 0 & (x,y) & (z,-y) & 0 & 0 & L_{v} \\ 0 & 0 & (-x,-z) & (z,x) & 0 & L_{c1} \\ 0 & 0 & 0 & (-z,-y) & (z,y) & L_{r1} \\ 0 & 0 & 0 & 0 & (-z,-x) & L_{c1} \\ J_{r1} & J_{r1} & J_{v1} & J_{f1} & J_{r1} & 0 \end{bmatrix}.$$

Table 7.3 Task poses

Pose 1	Pose 2
$\begin{bmatrix} -0.433 & 0.866 & 0.250 & -478.11 \\ 0.750 & -0.500 & -0.433 & -128.11 \\ -0.500 & 0 & -0.866 & -256.22 \\ 0 & 0 & 0 & 1 \end{bmatrix}$	$\begin{bmatrix} 0.500 & 0.707 & -0.500 & -597.49 \\ 0.500 & -0.707 & -0.500 & -102.52 \\ -0.707 & 0 & -0.707 & -144.98 \\ 0 & 0 & 0 & 1 \end{bmatrix}$
Pose 3	Pose 4
$\begin{bmatrix} 0.268 & 0.500 & -0.824 & -751.57 \\ 0.155 & -0.866 & -0.475 & -29.76 \\ -0.951 & 0 & -0.309 & 133.68 \\ 0 & 0 & 0 & 1 \end{bmatrix}$	$\begin{bmatrix} 0 & 0 & -1 & -700 \\ 0 & -1 & 0 & 350 \\ -1 & 0 & 0 & 350 \\ 0 & 0 & 0 & 1 \end{bmatrix}$

Table 7.4 The adaptive probabilities for the mutation operators (pose problem)

Average fitness value F_a	Link mutation operator	Port mutation operator
20	0.2	0.2
$10 \leq F_a < 20$	0.06	0.045
$5 \leq F_a < 10$	0.045	0.03
$3 \leq F_a < 5$	0.035	0.02
$F_a < 3$	0.03	0.015

Remark 1 In this section, we present two computation examples for demonstration purpose. The computation results are quite encouraging because the optimal robot configurations in both "position" and "pose" design examples are kinematically identical to the configurations used to generate the given task points and poses, respectively. We can also geometrically show that each of the derived optimal MDOF robots (2 DOFs) contains the minimal DOFs. Taking the "position" design example for instance, four given task points (Table 7.1) do not fall in a circle or a straight line. Hence, it would require at least two DOFs to perform this task.

Remark 2 The computation results also show that the reachability constraint has significant effect on the search efficiency. In the position design example, the first feasible configuration is derived in the second generation and the optimal configuration first appears in the 11th generation as shown in Fig. 7.10. In the pose design example, however, the first feasible configuration is derived in the 22nd generation and the optimal configuration first appears in the 47th generation as shown in Fig. 7.17. This is due to fact that the positional reachability constraint is easier to be satisfied. The pose design example takes about 6 h computation time on a SUN ULTRASPARC 1 workstation, while the position design example takes only 2 h computation time under the same environment.

Remark 3 It is the consideration of virtual modules that allows the algorithm to perform both type and dimension synthesis uniformly. To a certain extent, the search for optimal robot configuration is a procedure to increase the number of virtual and cubic link modules in the candidate configurations.

Remark 4 In the EA, there are several user-defined control parameters, such as the population size, mutation and crossover operators, which have significant effect on the search efficiency. Many trial computations have been done in order to draw the suitable values to be used for the control parameters. (1) The suitable population size is between 15 to 40; (2) the link and port mutation operators are between 0.01 to 0.06; (3) the crossover operator has little effect on the search efficiency and the recommended range is between 0.2 to 0.4. The algorithm, as demonstrated by the provided examples, is reliable and effective under the suitable control parameters.

Chapter 8
Modular Parallel Robot Kinematics

Abstract This chapter focuses on the kinematics analysis issues for a class of three-legged modular 6-DOF parallel robots, including displacement analysis, instantaneous kinematics analysis, singularity analysis and workspace analysis. Three approaches, i.e., sensor-based solution approach, numerical solution approach and projection approach, are proposed for the forward displacement analysis. Based on the Product-of-Exponential (POE) formula, a geometric formulation approach is employed for the instantaneous kinematics analysis. A passive-joint velocity based approach is proposed for the singularity analysis. Three parameterization methods and their corresponding partition schemes for spatial rigid-body rotations are discussed for the constant position workspace analysis.

8.1 Introduction

A Parallel Kinematic Machine (PKM) is a closed-loop mechanism in which the moving platform is connected to the base by at least two serial kinematic chains (legs). Compared with the series mechanisms, the parallel mechanisms have some particular features, such as high dynamic performance, great structural stiffness, high speed and high precision. Parallel mechanisms can be divided into 6-DOF robots and less DOF robots according to the number of DOFs. The conventional Stewart-Gough platform [151] has six extensible legs and hence leads to a rigid kinematic structure. However, such a six-leg fully parallel manipulator has coupled-motion characteristics, complicated kinematics, and limited workspace, which makes its design and application development difficult.

To overcome the difficulties in the traditional PKMs, the PKMs with a fewer number of legs, decoupled motions, simple kinematics, and large workspace have been received a great attention in recent robotics literature, e.g., the three-leg 3-DOF Universal Cartesian Robot (UCR) [152, 153], the three-leg 6-DOF Tri-Scott manipulator [154], the three-legged 6-DOF Eclipse manipulator [155], the three-leg

G. Yang and I.-M. Chen, *Modular Robots: Theory and Practice*, Research on Intelligent Manufacturing,
https://doi.org/10.1007/978-981-16-5007-9_8

6-DOF ReSl-Bot manipulator [156], the 3UPS PKM with three composite legs [157], and other RRRS, RRSR manipulators with rotating links [158]. It is noted that a PKM is considered having Decoupled-Motion Architecture (DMA) if some of its primary motions (e.g., translation along with or rotation about an axis) are always driven by certain (not all) active joints. A PKM with DMA also implies that the PKM should have simple kinematics so as to facilitate its analysis, design, and motion control.

Another major concern of PKMs with DMA is to investigate and formulate effective kinematics algorithms, such as displacement, singularity, and workspace analysis. These algorithms are essential for the performance analysis and design optimization of the PKM.

A number of displacement analysis approaches have been proposed for the three-legged non-redundant PKMs. Innocenti and Castelli [159] proposed an analytical method for the forward displacement analysis of PKMs, which needed to solve a 16th-order polynomial equation. Utilizing the sensed displacements of some passive joints, Notash and Podhorodeski [160] introduced a simplified analytical method for on-line computation of the moving platform poses. To avoid the complexity in solving high-order polynomial equations, Cleary and Brook [161] proposed a numerical method for the forward displacement analysis in which only one solution could be determined each time.

One of the early works to address configuration singularities of general closed-loop mechanisms was presented by Gosselin and Angeles [162], in which three types of configuration singularities were classified. Hunt [163] and Kumar [164] provided a general framework for applying screw theory to singularity analysis. Merlet [165] showed that it is possible to enumerate all the singularities of the Stewart-Gough platform by applying line geometry. Zhou [166] analyzed the Stewart manipulators for special orientations. Kamyar made a detection of complex singularities for a function of several variables [167]. Md. Shahidul proposed a boundary element method to solve the analysis of intensity of singularity [168].

A 6-DOF PKM, in general, has a six-dimensional (6-D) (position and orientation) workspace. As it is very difficult to present the complete 6-D workspace in a human readable way, different subsets of the complete workspace are usually determined such as the constant orientation workspace [169], the projected orientation workspace, and the orientation workspace [170]. A simple and generic approach to determine these workspace subsets was to use a discretization method [171–174]. An advanced geometric approach was first introduced by Gosselin [169], where boundaries of the horizontal cross sections of the workspace were determined. Merlet [175] extended this geometric approach by considering the joint limitations and leg interference. In recent years, some new methods to analyze the robot workspace have been proposed. Liu [176] presented a reconstructing solving method based on image engine. Zhao [177] combined an improved Monte Carlo method and voxel algorithm for the workspace analysis of a 9-DOF hyper-redundant manipulator. Adrián showed a method based on the vanishing of self-motion manifolds to determine the collision-free workspace of redundant robots [178]. The methods of analyzing the workspace generally rely on intelligent algorithms and machine learning, and the results will be more accurate and practical compared with traditional methods.

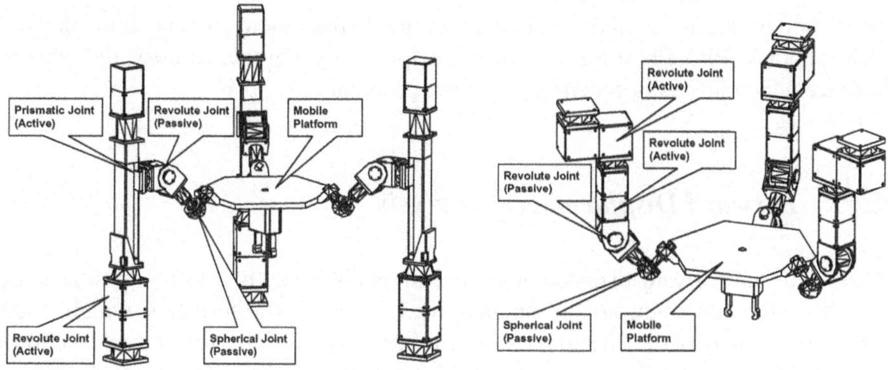

Fig. 8.1 Two modular three-leg parallel robot configurations

In this chapter, a class of three-legged modular 6-DOF parallel mechanisms is identified as having simple kinematics and desirable characteristics. Such a parallel robot consists of three legs, and each leg has two actuating joints, one passive revolute joint, and one passive spherical joint. Figure 8.1 shows two such possible three-legged parallel robot configurations, i.e., a three-legged RRRS-type parallel robot and a three-legged RRRS-type parallel robot. This chapter presents the forward kinematics position analysis approaches, i.e., the sensor-based solution approach, the numerical solution approach and the projection approach. The instantaneous kinematics analysis is derived based on passive joint velocities method. The singularity analyses are discussed for 3RRRS, 3RPRS PKMs by passive-joint-velocity based method and projection method, respectively. Finally, the constant orientation workspace analysis is presented for a 3RPRS PKM, and the numerical orientation workspace analysis with different parameterization method is proposed to achieve the visualization of 3-D orientation workspace, which is used to analyse and validate the workspace of a 3-DOF SPM.

8.2 Displacement Analysis

The purpose of displacement analysis for a parallel robot is to derive the kinematic relationship between the active joint displacements and the end-effector poses. It is divided into the forward displacement analysis and the inverse displacement analysis, the former is to determine the pose of the end-effector when the active joint displacements are known, and the latter is to determine the active joint displacements when the pose of the end-effector is given. For a PKM, the forward displacement analysis is a difficult analytical problem. It involves non-linear equations, and many solutions exist. Therefore, it is difficult to find the general solution algorithm for the forward displacement analysis of the parallel robots expect for some special parallel robot configurations. In this section, the forward displacement analyses for a 3RRRS PKM and a 3RPRS PKM are conducted, respectively. On the contrary, the inverse displacement analysis is generally an easier problem.

8.2.1 Forward Displacement Analysis

In general, the forward displacement analysis mainly has two forms of solutions, i.e., the numerical solution and the analytical solution. The former is widely used because it is easily obtained and possesses a simple form, but the result is not accurate enough. Instead, the latter has the complex analytic form.

The schematic diagram of the three-legged 6-DOF parallel robot configurations is shown in Fig. 8.2. We assume that joint ij ($\hat{s}_{ij}$) is the actuating joint ($i = 1, 2, 3$; $j = 1, 2$), and joint $i3$ ($\hat{s}_{i3}$) is the passive revolute joint ($i = 1, 2, 3$). Define A as the local frame attached to the mobile platform and B as the base frame. The coordinates of the passive spherical joint, point A_i, relative to frame A and the passive revolute joint frame B_{i3} are given by $p_i^{'} = \left(x_{ai}^{'}, y_{ai}^{'}, z_{ai}^{'}\right)^T$ and $p_i^{''} = \left(x_{ai}^{''}, y_{ai}^{''}, z_{ai}^{''}\right)^T$, respectively. The forward displacement analysis becomes to determine the pose of frame A with respect to the base frame B when the joint displacements of the six active joints, q_{ij}, are known.

Sensor-based Solution Approach

The sensor-based method is a simple and practical approach for the forward displacement analysis of parallel robots. The basic idea is to install sensors in the passive joint modules to measure the corresponding joint displacements.

As shown in Fig. 8.2, the position of point A_i with respect to the base frame B can be directly determined as a function of the actuating joint displacements and the sensed passive joint displacement in leg i. For each leg, the computation is identical to the forward kinematics of serial chains. Based on the local Product-of-Exponential (POE) formula (Eq. (4.30)), the position vector of point A_i, p_i, can be given by

$$\begin{bmatrix} p_i \\ 1 \end{bmatrix} = T_{B,B_{i0}} T_{B_{i0},B_{i1}}(0) e^{\hat{s}_{i1} q_{i1}} T_{B_{i1},B_{i2}}(0) e^{\hat{s}_{i2} q_{i2}} T_{B_{i2},B_{i3}}(0) e^{\hat{s}_{i3} q_{i3}} \begin{bmatrix} p_i^{''} \\ 1 \end{bmatrix}, \tag{8.1}$$

where $T_{B,B_{i0}}$ is the fixed kinematic transformation from frame B to frame B_{i0} (the base frame of leg i).

Note that the homogeneous coordinate representation is employed in Eq. (8.1). Denote the forward kinematic transformation from frame B to frame A as $T_{B,A}$. Then

$$T_{B,A} \begin{bmatrix} p_i^{'} \\ 1 \end{bmatrix} = \begin{bmatrix} p_i \\ 1 \end{bmatrix}. \tag{8.2}$$

We also have

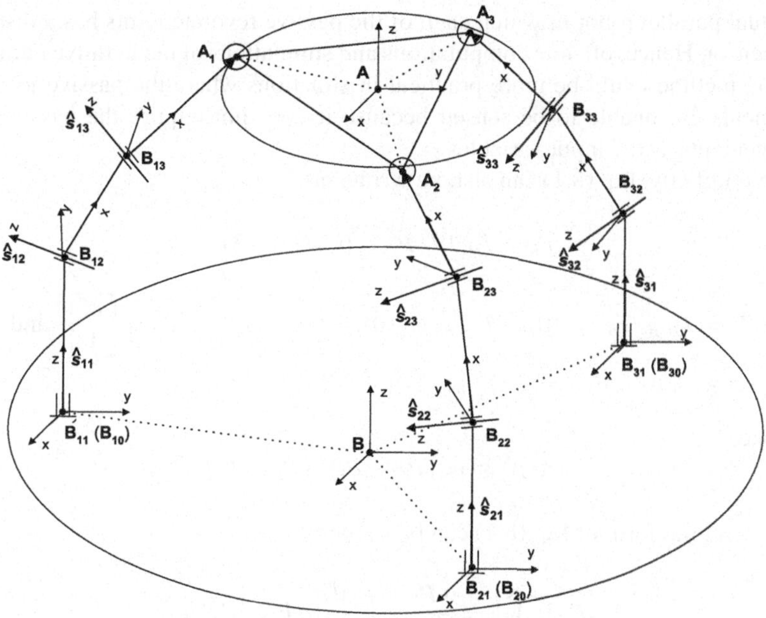

Fig. 8.2 Kinematic structure of a three-legged parallel robot

$$T_{B,A}\begin{bmatrix} p'_{12} \times p'_{23} \\ 0 \end{bmatrix} = \begin{bmatrix} p_{12} \times p_{23} \\ 0 \end{bmatrix}, \tag{8.3}$$

where $p'_{12} = p'_2 - p'_1$, $p'_{23} = p'_3 - p'_2$, $p_{12} = p_2 - p_1$, and $p_{23} = p_3 - p_2$.

Combining Eqs. (8.2) and (8.3), we obtain

$$T_{B,A}\begin{bmatrix} p'_1 & p'_2 & p'_3 & p'_{12} \times p'_{23} \\ 1 & 1 & 1 & 0 \end{bmatrix} = \begin{bmatrix} p_1 & p_2 & p_3 & p_{12} \times p_{23} \\ 1 & 1 & 1 & 0 \end{bmatrix}. \tag{8.4}$$

Therefore, the pose of the mobile platform, $T_{B,A}$, can be given by

$$T_{B,A} = \begin{bmatrix} p_1 & p_2 & p_3 & p_{12} \times p_{23} \\ 1 & 1 & 1 & 0 \end{bmatrix}\begin{bmatrix} p'_1 & p'_2 & p'_3 & p'_{12} \times p'_{23} \\ 1 & 1 & 1 & 0 \end{bmatrix}^{-1}. \tag{8.5}$$

Note that the matrix $\begin{bmatrix} p'_1 & p'_2 & p'_3 & p'_{12} \times p'_{23} \\ 1 & 1 & 1 & 0 \end{bmatrix}$ is a constant matrix for a special robot configuration, and it is always invertible if point A_1, A_2 and A_3 neither coincide with each other nor fall on the same line.

Numerical Solution Approach

The limitation of the sensor-based algorithm is that it can only be implemented on the actual parallel robot in which each of the passive revolute joints has a displacement sensor. Hence, off-line computations and simulations of the iterative numerical solution method could be more practical in situations where the passive joint displacements are unable to be sensed because it does not require the passive joint displacements as the input variables.

For simplicity, Eq. (8.1) can also be written as

$$P_i = T_i e^{\hat{s}_{i3} q_{i3}} P_i'' \quad (i = 1, 2, 3), \tag{8.6}$$

where $T_i = T_{B,B_{i0}} T_{B_{i0},B_{i1}}(0) e^{\hat{s}_{i1} q_{i1}} T_{B_{i1},B_{i2}}(0) e^{\hat{s}_{i2} q_{i2}} T_{B_{i2},B_{i3}}(0)$, $P_i = \begin{bmatrix} p_i \\ 1 \end{bmatrix}$, and $P_i'' = \begin{bmatrix} p_i'' \\ 1 \end{bmatrix}$.

Since

$$\|\overrightarrow{A_1 A_2}\|^2 = (P_2 - P_1)^T (P_2 - P_1). \tag{8.7}$$

The differential form of Eq. (8.7) can be written as

$$d\|\overrightarrow{A_1 A_2}\| = \frac{(P_2 - P_1)^T}{\|\overrightarrow{A_1 A_2}\|} (dP_2 - dP_1). \tag{8.8}$$

According to Eq. (8.6), we find

$$dP_i = T_i e^{\hat{s}_{i3} q_{i3}} \hat{s}_{i3} P_i'' dq_{i3}. \tag{8.9}$$

The actual distance from point A_1 to point A_2 is a constant, denoted by a_{12}. Then,

$$d\|\overrightarrow{A_1 A_2}\| = a_{12} - \|\overrightarrow{A_1 A_2}\| = \frac{(P_2 - P_1)^T}{\|\overrightarrow{A_1 A_2}\|} (dP_2 - dP_1). \tag{8.10}$$

Similarly, we have

$$a_{23} - \|\overrightarrow{A_2 A_3}\| = \frac{(P_3 - P_2)^T}{\|\overrightarrow{A_2 A_3}\|} (dP_3 - dP_2), \tag{8.11}$$

$$a_{31} - \|\overrightarrow{A_3 A_1}\| = \frac{(P_1 - P_3)^T}{\|\overrightarrow{A_3 A_1}\|} (dP_1 - dP_3). \tag{8.12}$$

Combining Eqs. (8.10), (8.11), and (8.12), we get

$$dq = J^{-1} da, \tag{8.13}$$

where

$$dq = column\,[dq_{13}, dq_{23}, dq_{33}]\,,$$
$$da = column\left[a_{12} - \|\overrightarrow{A_1A_2}\|, a_{23} - \|\overrightarrow{A_2A_3}\|, a_{31} - \|\overrightarrow{A_3A_1}\|\right],$$
$$J = \begin{bmatrix} -\frac{(P_2-P_1)^T}{\|\overrightarrow{A_1A_2}\|}\dot{P}_1 & \frac{(P_2-P_1)^T}{\|\overrightarrow{A_1A_2}\|}\dot{P}_2 & 0 \\ 0 & -\frac{(P_3-P_2)^T}{\|\overrightarrow{A_2A_3}\|}\dot{P}_2 & \frac{(P_3-P_2)^T}{\|\overrightarrow{A_2A_3}\|}\dot{P}_3 \\ \frac{(P_1-P_3)^T}{\|\overrightarrow{A_3A_1}\|}\dot{P}_1 & 0 & -\frac{(P_1-P_3)^T}{\|\overrightarrow{A_3A_1}\|}\dot{P}_3 \end{bmatrix},$$

in which $\dot{P}_i$ can be obtained according to Eq. (8.9).

Equation (8.13) can be written as an iterative form, i.e.,

$$q^{(k+1)} = q^{(k)} + (J^{-1}da)^{(k)}, \tag{8.14}$$

where k represents the number of iterations. Based on the standard iterative form of Eq. (8.14), the Newton Raphson method is employed to derive the numerical solution of $q(= (q_{13}, q_{23}, q_{33})^T)$. Since the Jacobian matrix, J, is a 3×3 matrix, the inverse of the Jacobian matrix J can be found in a symbolic form. From computational point of view, however, the proposed algorithm is equivalent to the method given by Cleary and Brooks [161]. After the joint displacement q is derived, the pose of the mobile platform can be easily determined by using the sensor-based algorithm.

Projection Approach

Due to the special kinematic features of the 3RPRS PKM with DMA such that all joint axes, except for the three spherical joints at the leg ends, are parallel to each other and perpendicular to the base plane, a projection approach can be applied in order to further simplify the forward displacement analysis [183, 184]. The schematic diagram of the 3RPRS PKM with DMA is shown in Fig. 8.3. By projecting the 3RPRS PKM onto the base (horizontal) plane, an imaginary 3RRRS planar PKM can be obtained as shown in Fig. 8.4, where the prismatic joints are temporarily ignored and the spherical joints are treated as revolute joints. On the other hand, the linear motions of the three prismatic joints can be directly determined when projecting the 3RPRS PKM onto a vertical direction.

The forward displacement analysis of such a general 3RRR planar PKM has been well studied by Gosselin and Sefrioui [189], in which a closed-form polynomial solution was derived. Since the closed-form solution leads to a polynomial of degree six, at most six real solutions exist for the projected 3RRR PKM. However, since the 3RRR PKM is a spatial mechanism, there are two possible triangular moving plates: the original plate and its mirror, satisfying the same three edge-length constraints [183]. Therefore, we can conclude that at most 12 real solutions exist for the forward displacement analysis of the 3RPRS PKM with DMA. For more details about this method, please refer to [183].

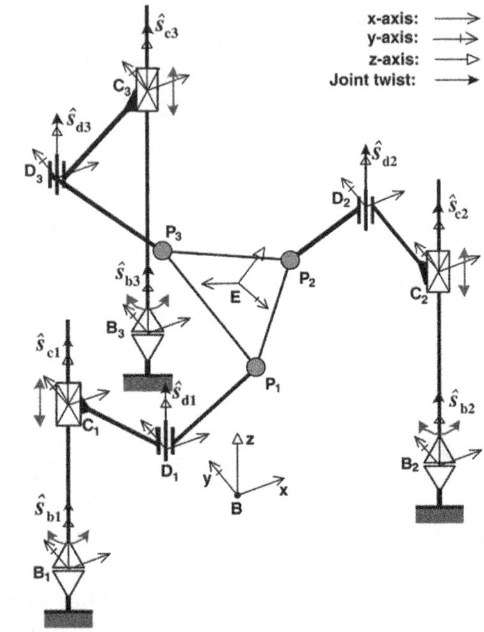

Fig. 8.3 A Schematic diagram of the 3RPRS PKM with DMA

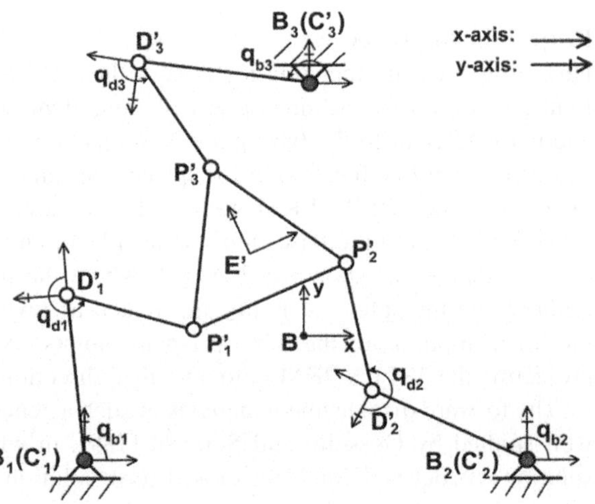

Fig. 8.4 Projected planar 3RRR PKM

Computation Example

In order to demonstrate the effectiveness of the proposed forward displacement analysis algorithms for a modular three-leg 6-DOF parallel robot as shown in Fig. 8.2, a computation example is given. The given geometric parameters are as follows:

$$T_{B,B_{10}} = \begin{bmatrix} 0.5 & -0.866 & 0 & -250 \\ 0.866 & 0.5 & 0 & -433 \\ 0 & 0 & 1 & 80 \\ 0 & 0 & 0 & 1 \end{bmatrix},$$

$$T_{B,B_{20}} = \begin{bmatrix} -1 & 0 & 0 & 500 \\ 0 & -1 & 0 & 0 \\ 0 & 0 & 1 & 80 \\ 0 & 0 & 0 & 1 \end{bmatrix},$$

$$T_{B,B_{30}} = \begin{bmatrix} 0.5 & 0.866 & 0 & -250 \\ -0.866 & 0.5 & 0 & 433 \\ 0 & 0 & 1 & 80 \\ 0 & 0 & 0 & 1 \end{bmatrix},$$

$$T_{B_{10},B_{11}}(0) = T_{B_{20},B_{21}}(0) = T_{B_{30},B_{31}}(0) = \begin{bmatrix} 0 & 0 & 1 & 90 \\ 0 & -1 & 0 & 0 \\ 1 & 0 & 0 & 0 \\ 0 & 0 & 0 & 1 \end{bmatrix},$$

$$T_{B_{11},B_{12}}(0) = T_{B_{21},B_{22}}(0) = T_{B_{31},B_{32}}(0) = \begin{bmatrix} 0 & 0 & 1 & 195 \\ 0 & 1 & 0 & 0 \\ -1 & 0 & 0 & 0 \\ 0 & 0 & 0 & 1 \end{bmatrix},$$

$$T_{B_{12},B_{13}}(0) = T_{B_{22},B_{23}}(0) = T_{B_{32},B_{33}}(0) = \begin{bmatrix} 1 & 0 & 0 & 0 \\ 0 & 0 & -1 & 0 \\ 0 & 1 & 0 & 105 \\ 0 & 0 & 0 & 1 \end{bmatrix},$$

$s_{ij} = (0, 0, 0, 0, 0, 1)^T (i, j = 1, 2, 3)$,
$p_1' = (-150, -259.81, 0)^T$, $p_2' = (300, 0, 0)^T$, $p_3' = (-150, 259.81, 0)^T$,
$p_1'' = p_2'' = p_3'' = (0, 500, 0)^T$.

The entire computation scheme consists of two parts. The first part is to solve for the displacements of passive joints $q_p = (q_{13}, q_{23}, q_{33})^T$ with the known actuating joint displacements $q_a = (q_{11}, q_{21}, q_{31}, q_{12}, q_{22}, q_{32})^T$ by using the iterative numerical algorithm. The second part, based on the derived solution of q_p, is to determine the pose of the mobile platform $T_{B,A}$ by using the sensor-based algorithm. The computation results are listed in Table 8.1. In the right column: q_a—the input actuating joint displacements; q_p—the derived passive joint displacements; and N—the number of iterations. The left column represents the derived poses of the mobile platform $T_{B,A}$. The initial guess solutions for the passive joints, q_p, are identically taken as $(0, 0, 0)^T$. The desired solutions in all five cases can be obtained in 3 to 4 iterations. Nevertheless, the iterative numerical algorithm cannot guarantee that a solution will be derived within a fixed number of iterations. The computation efficiency depends on the initial guess solutions. Note that in the computation examples, the joint angles are expressed in radians (for revolute joints), and the position vectors are in millimeters.

Table 8.1 Computation results

Input	Derived poses
$q_a = (0, 0, 0, 0, 0, 0)^T$ $q_p = (0.2218, 0.2218, 0.2218)^T$ $N = 3$	$\begin{bmatrix} 1 & 0 & 0 & 0 \\ 0 & 1 & 0 & 0 \\ 0 & 0 & 1 & 867.75 \\ 0 & 0 & 0 & 1 \end{bmatrix}$
$q_a = (0, 0.1, 0.2, 0, 0.1, 0.2)^T$ $q_p = (0.1386, 0.4121, 0.1735)^T$ $N = 3$	$\begin{bmatrix} 0.9737 & -0.2195 & 0.0616 & -81.42 \\ 0.2170 & 0.9752 & 0.0436 & -9.32 \\ -0.0696 & -0.0291 & 0.9972 & 857.21 \\ 0 & 0 & 0 & 1 \end{bmatrix}$
$q_a = (0, 0.2, 0.4, 0, 0.2, 0.4)^T$ $q_p = (0.0605, 0.6486, 0.1882)^T$ $N = 4$	$\begin{bmatrix} 0.9084 & -0.4038 & 0.1089 & -158.52 \\ 0.3840 & 0.9085 & 0.1651 & -35.25 \\ -0.1656 & -0.1081 & 0.9802 & 826.15 \\ 0 & 0 & 0 & 1 \end{bmatrix}$
$q_a = (0, 0.3, 0.6, 0, 0.3, 0.6)^T$ $q_p = (0.0011, 0.9069, 0.2650)^T$ $N = 4$	$\begin{bmatrix} 0.8367 & -0.5313 & 0.1328 & -217.24 \\ 0.4673 & 0.8191 & 0.3327 & -71.67 \\ -0.2855 & -0.2163 & 0.9336 & 780.98 \\ 0 & 0 & 0 & 1 \end{bmatrix}$
$q_a = (0, 0.4, 0.8, 0, 0.4, 0.8)^T$ $q_p = (-0.0280, 1.1732, 0.4047)^T$ $N = 4$	$\begin{bmatrix} 0.7843 & -0.5987 & 0.1623 & -249.90 \\ 0.4532 & 0.7317 & 0.5091 & -109.10 \\ -0.4236 & -0.3258 & 0.8452 & 731.63 \\ 0 & 0 & 0 & 1 \end{bmatrix}$

8.2.2 *Inverse Displacement Analysis*

The inverse displacement analysis is to determine the six active joint displacements when the pose of the mobile platform frame with respect to the base frame is given.

As shown in Fig. 8.2, when the pose of the mobile platform frame $T_{B,A}$ is given, the position vector of point A_i with respect to the base frame, p_i, can be directly determined according to Eq. (8.2). Once p_i is known, each leg can be treated as an individual serial manipulator and any inverse kinematics algorithm for serial-type manipulators can be applied to complete the analysis. Note that the inverse kinematics solution of the parallel manipulator exists if and only if the inverse kinematics solutions exist for every individual leg. Following the POE approach, we use the Paden–Kahan method for the inverse kinematics analysis.

8.3 Instantaneous Kinematics Analysis

The instantaneous kinematics analysis pertains to the study of the velocity relationships among the active joints, the passive joints, and the mobile platform. Such kinematics relationships, obviously, will mainly be used for the singularity analysis and dynamic analysis.

8.3.1 Forward Instantaneous Kinematics Analysis

The purpose of the forward instantaneous kinematics analysis is to determine the passive joint velocities and then the mobile platform velocity from the actuator joint velocities. For the 3RRRS PKM, as shown in Fig. 8.2, using the base frame representation of leg-end point A_i, i.e., $p_i \in \Re^{3\times1}$ $(i = 1, 2, 3)$, the three constant leg-end distances can be written as

$$\begin{aligned}
\|\overrightarrow{A_1A_2}\|^2 &= (p_2 - p_1)^T(p_2 - p_1),\\
\|\overrightarrow{A_2A_3}\|^2 &= (p_3 - p_2)^T(p_3 - p_2),\\
\|\overrightarrow{A_3A_1}\|^2 &= (p_1 - p_3)^T(p_1 - p_3).
\end{aligned} \tag{8.15}$$

Differentiating both sides of Eq. (8.15) with respect to time, we have

$$\begin{aligned}
(p_2 - p_1)^T(\dot{p}_2 - \dot{p}_1) &= 0,\\
(p_3 - p_2)^T(\dot{p}_3 - \dot{p}_2) &= 0,\\
(p_1 - p_3)^T(\dot{p}_1 - \dot{p}_3) &= 0,
\end{aligned} \tag{8.16}$$

where $\dot{p}_i$ is the leg-end velocity of leg i.

Differentiating both sides of Eq. (8.1) with respect to time, we have

$$\begin{aligned}
\begin{bmatrix}\dot{p}_i\\0\end{bmatrix} &= T_{B,B_{i0}}T_{B_{i0},B_{i1}}(0)e^{\hat{s}_{i1}q_{i1}}\hat{s}_{i1}T_{B_{i1},B_{i2}}(0)e^{\hat{s}_{i2}q_{i2}}T_{B_{i2},B_{i3}}(0)e^{\hat{s}_{i3}q_{i3}}\begin{bmatrix}p_i''\\1\end{bmatrix}\dot{q}_{i1}\\
&+ T_{B,B_{i0}}T_{B_{i0},B_{i1}}(0)e^{\hat{s}_{i1}q_{i1}}T_{B_{i1},B_{i2}}(0)e^{\hat{s}_{i2}q_{i2}}\hat{s}_{i2}T_{B_{i2},B_{i3}}(0)e^{\hat{s}_{i3}q_{i3}}\begin{bmatrix}p_i''\\1\end{bmatrix}\dot{q}_{i2}\\
&+ T_{B,B_{i0}}T_{B_{i0},B_{i1}}(0)e^{\hat{s}_{i1}q_{i1}}T_{B_{i1},B_{i2}}(0)e^{\hat{s}_{i2}q_{i2}}T_{B_{i2},B_{i3}}(0)e^{\hat{s}_{i3}q_{i3}}\hat{s}_{i3}\begin{bmatrix}p_i''\\1\end{bmatrix}\dot{q}_{i3},
\end{aligned} \tag{8.17}$$

where $\dot{q}_{ij}$ is the velocity of joint ij $(j = 1, 2, 3)$. Let $T_{B,B_{ij}}$ represent the forward kinematic transformation from frame B to frame B_{ij} and $p'_{ij,i}$ (with $p'_{i3,i} = p'_i$) be the position vector of point A_i relative to frame B_{ij}. Equation (8.17) can be rewritten as:

$$\begin{aligned}
\begin{bmatrix}\dot{p}_i\\0\end{bmatrix} &= T_{B,B_{i1}}\hat{s}_{i1}\begin{bmatrix}p'_{i1,i}\\1\end{bmatrix}\dot{q}_{i1} + T_{B,B_{i2}}\hat{s}_{i2}\begin{bmatrix}p'_{i2,i}\\1\end{bmatrix}\dot{q}_{i2}\\
&+ T_{B,B_{i3}}\hat{s}_{i3}\begin{bmatrix}p'_{i3,i}\\1\end{bmatrix}\dot{q}_{i3}\ .
\end{aligned} \tag{8.18}$$

Let $\hat{s}$ represent a twist, i.e., an element of $se(3)$, such that $\hat{s} = \begin{bmatrix} \hat{w} & v \\ 0 & 0 \end{bmatrix}$, and $p \in R^{3\times 1}$ be any position vector. We have

$$\hat{s}\begin{bmatrix} p \\ 1 \end{bmatrix} = \begin{bmatrix} \hat{w} & v \\ 0 & 0 \end{bmatrix}\begin{bmatrix} p \\ 1 \end{bmatrix} = \begin{bmatrix} I_{3\times 3} & -\hat{p} \\ 0 & 0 \end{bmatrix}\begin{bmatrix} v \\ w \end{bmatrix}, \tag{8.19}$$

where $\hat{p} \in \Re^{3\times 3}$ is the cross-product matrix (a skew-symmetric matrix) of vector $p \in \Re^{3\times 1}$. In Eq. (8.19), the block matrix $\begin{bmatrix} I_{3\times 3} & -\hat{p} \end{bmatrix} \in \Re^{3\times 6}$ can be considered as the transition matrix associated with the position vector p. Geometrically, it transmits the rigid body motion to a point motion defined by p. We term such a matrix as the position transition matrix and denote it as T_p. Equation (8.19) can then be rewritten as

$$\hat{s}\begin{bmatrix} p \\ 1 \end{bmatrix} = \begin{bmatrix} T_p \\ 0 \end{bmatrix} s, \tag{8.20}$$

where $s = (v, w) \in R^{6\times 1}$ is a 6−dimensional vector representation of $\hat{s}$. Substituting Eq. (8.20) into Eq. (8.18), we have

$$\begin{aligned} \begin{bmatrix} \dot{p}_i \\ 0 \end{bmatrix} &= T_{B,B_{i1}}\begin{bmatrix} T_{p'_{i1,i}} \\ 0 \end{bmatrix} s_{i1}\dot{q}_{i1} + T_{B,B_{i2}}\begin{bmatrix} T_{p'_{i2,i}} \\ 0 \end{bmatrix} s_{i2}\dot{q}_{i2} \\ &+ T_{B,B_{i3}}\begin{bmatrix} T_{p'_{i3,i}} \\ 0 \end{bmatrix} s_{i3}\dot{q}_{i3}, \end{aligned} \tag{8.21}$$

where s_{ij} is the twist coordinates of joint ij expressed in frame B_{ij}. Since

$$T_{B,B_{ij}} = \begin{bmatrix} R_{B,B_{ij}} & p_{B,B_{ij}} \\ 0 & 1 \end{bmatrix},$$

in which $R_{B,B_{ij}}$ and $p_{B,B_{ij}}$ represent the orientation and position of frame B_{ij} with respect to the base frame B, respectively. Equation (8.21) can also be further simplified as

$$\dot{p}_i = R_{B,B_{i1}} T_{p'_{i1,i}} s_{i1}\dot{q}_{i1} + R_{B,B_{i2}} T_{p'_{i2,i}} s_{i2}\dot{q}_{i2} + R_{B,B_{i3}} T_{p'_{i3,i}} s_{i3}\dot{q}_{i3}. \tag{8.22}$$

Kinematically, $R_{B,B_{ij}} T_{p'_{ij,i}} s_{ij}\dot{q}_{ij} \in R^{3\times 1}$ $(i, j = 1, 2, 3)$ represents the velocity of point A_i resulting from the motion of joint ij, i.e., $\dot{q}_{ij}$. Therefore, the term $R_{B,B_{ij}} T_{p'_{ij,i}}$ $s_{ij} \in R^{3\times 1}$ reflects the contribution of the ijth joint velocity to the point velocity of A_i. Denoting this term by u_{ij}, Eq. (8.22) becomes

$$\dot{p}_i = u_{i1}\dot{q}_{i1} + u_{i2}\dot{q}_{i2} + u_{i3}\dot{q}_{i3} \quad (i = 1, 2, 3). \tag{8.23}$$

Substituting Eq. (8.23) into Eq. (8.16), we have

$$J_p \dot{q}_p = J_a \dot{q}_a, \tag{8.24}$$

where $\dot{q}_p = [\dot{q}_{13}, \dot{q}_{23}, \dot{q}_{33}]^T \in \Re^{3\times1}$ represents the velocity of the passive joints, and $\dot{q}_a = [\dot{q}_{11}, \dot{q}_{12}, \dot{q}_{21}, \dot{q}_{22}, \dot{q}_{31}, \dot{q}_{32}]^T \in \Re^{6\times1}$ represents the velocity of the active joints. Moreover,

$$J_p = \begin{bmatrix} -p_{12}^T u_{13} & p_{12}^T u_{23} & 0 \\ 0 & -p_{23}^T u_{23} & p_{23}^T u_{33} \\ p_{31}^T u_{13} & 0 & -p_{31}^T u_{33} \end{bmatrix} \in \Re^{3\times3},$$

$$J_a = \begin{bmatrix} p_{12}^T u_{11} & p_{12}^T u_{12} & -p_{12}^T u_{21} & -p_{12}^T u_{22} & 0 & 0 \\ 0 & 0 & p_{23}^T u_{21} & p_{23}^T u_{22} & -p_{23}^T u_{31} & -p_{23}^T u_{32} \\ -p_{31}^T u_{11} & -p_{31}^T u_{12} & 0 & 0 & p_{31}^T u_{31} & p_{31}^T u_{32} \end{bmatrix} \in \Re^{3\times6},$$

$$p_{ij} = p_j - p_i.$$

If the Jacobian matrix J_p is non-singular, the passive-joint velocity vector $\dot{q}_p$ is given by

$$\dot{q}_p = J_p^{-1} J_a \dot{q}_a. \tag{8.25}$$

Since J_p is a 3×3 matrix with a simple structure, $J_p{}^{-1}$ can also be simply computed in symbolic form. Equation (8.25) shows that the passive-joint velocities can be determined from the active joint velocities. When both the active and passive joint velocities are all known, the mobile platform velocity can be readily derived as described below.

Differentiating both sides of Eq. (8.5) with respect to time, we have

$$\dot{T}_{B,A} = \begin{bmatrix} \dot{p}_1 & \dot{p}_2 & \dot{p}_3 & (\dot{p}_2 - \dot{p}_1) \times p_{23} + p_{12} \times (\dot{p}_3 - \dot{p}_2) \\ 0 & 0 & 0 & 0 \end{bmatrix} \cdot \begin{bmatrix} p_1'' & p_2'' & p_3'' & (p_2'' - p_1'') \times (p_3'' - p_2'') \\ 1 & 1 & 1 & 0 \end{bmatrix}^{-1}. \tag{8.26}$$

Right-multiplying both sides of Eq. (8.26) with $T_{B,A}^{-1}$, we obtain

$$\dot{T}_{B,A} T_{B,A}^{-1} = \begin{bmatrix} \dot{p}_1 & \dot{p}_2 & \dot{p}_3 & (\dot{p}_2 - \dot{p}_1) \times p_{23} + p_{12} \times (\dot{p}_3 - \dot{p}_2) \\ 0 & 0 & 0 & 0 \end{bmatrix} \cdot \begin{bmatrix} p_1 & p_2 & p_3 & p_{12} \times p_{23} \\ 1 & 1 & 1 & 0 \end{bmatrix}^{-1}. \tag{8.27}$$

Note that the matrix whose inverse appears in Eq. (8.27) is always invertible if points A_1, A_2, and A_3 neither coincident nor collinear, such that $p_{12} \times p_{23} \neq 0$.

In Eq. (8.27), the term, $\dot{T}_{B,A} T_{B,A}^{-1}$, represents the spatial velocity of the mobile platform frame A with respect to the base frame B, which is a twist. Denote the spatial velocity of the mobile platform $\dot{T}_{B,A} T_{B,A}^{-1}$ by $\hat{v}_{B,A}^s$, which has the form

$$\hat{\mathrm{v}}^s_{B,A} = \begin{bmatrix} \hat{w}^s_{B,A} & v^s_{B,A} \\ 0 & 0 \end{bmatrix} \in se(3), \tag{8.28}$$

where $w^s_{B,A}$ represents the instantaneous angular velocity of the mobile platform as viewed in the spatial (base) frame. However, $v^s_{B,A}$ is not the point velocity of the origin of the mobile platform frame with respect to the base frame. Rather, $v^s_{B,A}$ is the velocity of a point on the same rigid body of the mobile platform which coincides instantaneously with the origin of the base frame. In other words, if one stands at the origin of the base (spatial) frame and measures the instantaneous velocity of a point attached to the mobile platform and passing through the origin, one obtains $v^s_{B,A}$. A formal definition of the spatial velocity of a rigid body can be found in [62].

8.3.2 *Inverse Instantaneous Kinematics Analysis*

The purpose of the inverse instantaneous kinematics analysis is to determine the passive and active joint velocities from the mobile platform velocity. For the 3<u>RR</u>RS PKM, the velocity of the mobile platform is represented by its spatial velocity $\hat{\mathrm{v}}^s_{B,A}$. The velocity of point A_i $(i = 1, 2, 3)$ thus can be computed as

$$\begin{bmatrix} \dot{p}_i \\ 0 \end{bmatrix} = \hat{\mathrm{v}}^s_{B,A} \begin{bmatrix} p_i \\ 1 \end{bmatrix}. \tag{8.29}$$

Based on Eq. (8.20), Eq. (8.29) has the form

$$\dot{p}_i = T_{p_i} \mathrm{v}^s_{A,B}, \tag{8.30}$$

where $\mathrm{v}^s_{B,A} = (v^s_{B,A}, w^s_{B,A})^T \in \Re^{6\times1}$ is the 6-dimensional vector representation of $\hat{\mathrm{v}}^s_{B,A}$ and $T_{p_i} \in \Re^{3\times6}$ is the position transition matrix associated with p_i, i.e., $T_{p_i} = \begin{bmatrix} I_{3\times3} & -\hat{p}_i \end{bmatrix}$.

Equation (8.23) can also be rewritten as:

$$\dot{p}_i = J_i \dot{q}_i, \tag{8.31}$$

where $\dot{q}_i = [\dot{q}_{i1}, \dot{q}_{i2}, \dot{q}_{i3}]^T \in \Re^{3\times1}$ represents the (passive and active) joint velocity vector of leg i, and $J_i = [u_{i1} \;\; u_{i2} \;\; u_{i3}] \in \Re^{3\times3}$ represents the Jacobian matrix of leg i.

For the analysis at hand, the three legs are decoupled from each other. If the Jacobian matrix of leg i, J_i, is non-singular, then the joint velocity vector of leg i, $\dot{q}_i$, is given by

$$\dot{q}_i = J_i^{-1} \dot{p}_i = J_i^{-1} T_{p_i} \mathrm{v}^s_{A,B}. \tag{8.32}$$

Using Eq. (8.32), both the passive and active joint velocities can be simultaneously determined from the mobile platform velocity.

Based on the POE formula, both forward and inverse kinematics analysis algorithms have been formulated. It is noted that the kinematic equations derived from the POE formula have compact forms and well-defined structures such that they can be readily differentiated with respect to time. It has been demonstrated, through a number of algorithms, that the POE formulation approach is also computationally efficient [87]. All these features make the POE formula an effective tool for the formulation of the instantaneous kinematics models.

8.4 Singularity Analysis

One of the main issues in the parallel robots is the kinematic singularity analysis. The notion of singularity in parallel robot kinematics refers to the configurations in which a robot either loses or gains one or more DOFs instantaneously. In other words, a modular parallel robot will lose its designated motion and working capability in a singular configuration. Hence, to design a parallel robot with a desired performance, e.g., high rigidity and manipulability, singularities must be excluded from its working area, if possible. This section mainly focuses on the singularity analysis of a class of three-legged 6-DOF modular parallel robots. Based on the singularities of matrices J_p and J_i derived in Sect. 8.3, we classify the singularities of the three-legged parallel robots into three categories, i.e., the forward, inverse, and combined singularities. Different from the conventional singularity analysis methods, two methods are proposed to simplify the singularity analysis.

The first method is based on passive-joint velocities. If the passive-joint velocities as well as the mobile platform velocity can be determined from the active-joint velocities, there will be no forward singularity. On the other hand, if the passive- and active-joint velocities can be determined from the mobile platform velocity, there will be no inverse singularity. Excluding the passive spherical joints at the leg ends, there are only three 1-DOF passive joints in a three-legged 6-DOF parallel robot. The complexity of the singularity analysis is significantly reduced because only four 3×3 matrices need to be analyzed.

The other is a geometric projection approach specifically proposed for 3RPRS PKM. Because the classical approach, i.e., rank analysis of the Jacobian matrices, usually is not an easy task for a 6-DOF PKM, the simple geometric approach is employed to identify the singularity conditions. Formulation of instantaneous kinematics and Jacobian matrices can be avoided. Note that the mechanical interferences are ignored in the singularity analysis.

8.4.1 Forward Singularity

The forward singularity configuration refers to a specific robot configuration in which the moving platform gains one or more degrees of freedom, instantaneously. Kinematically, it also means that there exists some nonzero vectors of the moving platform velocity when all of the active-joint velocities are equal to zero. In other words, if all of the active joints are completely locked, the moving platform will still posses infinitesimal motion in certain directions. Examples utilizing the proposed methods are presented as follows.

Passive-joint-velocity based method

From the instantaneous velocity analysis in Sect. 8.3, we realize that a singularity will occur when J_p or any J_i $(i = 1, 2, 3)$ becomes a singular matrix. The forward singularity occurs when matrix $J_p \in \Re^{3\times3}$ is singular. If J_p is singular, we will not be able to determine the passive joint velocities as well as the mobile platform velocity from the active-joint velocities. Matrix J_p has the form

$$J_p = \begin{bmatrix} -p_{12}^T u_{13} & p_{12}^T u_{23} & 0 \\ 0 & -p_{23}^T u_{23} & p_{23}^T u_{33} \\ p_{31}^T u_{13} & 0 & -p_{31}^T u_{33} \end{bmatrix} \in \Re^{3\times3}. \tag{8.33}$$

The determinant Δ_p of matrix J_p can be computed as

$$\Delta_p = (p_{12}^T u_{23})(p_{23}^T u_{33})(p_{31}^T u_{13}) - (p_{12}^T u_{13})(p_{23}^T u_{23})(p_{31}^T u_{33}). \tag{8.34}$$

Obviously, if $\Delta_p = 0$, a forward singularity will occur. In principle, Δ_p can be numerically determined from Eq. (8.34) when the active- and passive-joint angles are known. Since the passive joints are usually revolute joints and vectors p_{12}, p_{23}, and p_{31} cannot be zero, we can obtain the following three geometric conditions for the forward singularities:

- A forward singularity will occur if any passive-joint axis $\hat{s}_{i3}$ passes through its corresponding leg-end point A_i $(i = 1, 2, 3)$. It follows that the contribution of the i3th joint velocity to the point velocity of A_i vanishes, i.e., $u_{i3} = 0$. At least one of the column vectors in J_p (Eq. (8.33)) vanishes and hence, Δ_p=0.
  Example 1: Figure 8.5 shows a particular three-legged parallel robot in which the three passive joint axes $\hat{s}_{13}$, $\hat{s}_{23}$, and $\hat{s}_{33}$ always pass through their corresponding leg-end points A_1, A_2, and A_3. As a result, $u_{13} = u_{23} = u_{33} = 0$ and $\Delta_p = 0$. In fact, this forward singularity is caused by a particular robot architecture and exists for all configurations inside the entire robot workspace. This is, therefore, an architecture singularity [190] and can be avoided by design.
- A forward singularity will occur if any passive joint axis $\hat{s}_{i3}$ falls in the plane formed by points A_1, A_2, and A_3. It follows that u_{i3} will be perpendicular to the plane as well as any vector in this plane, e.g., p_{12}, p_{23}, and p_{31}. At least one of the column vectors in J_p vanishes and hence, Δ_p=0.

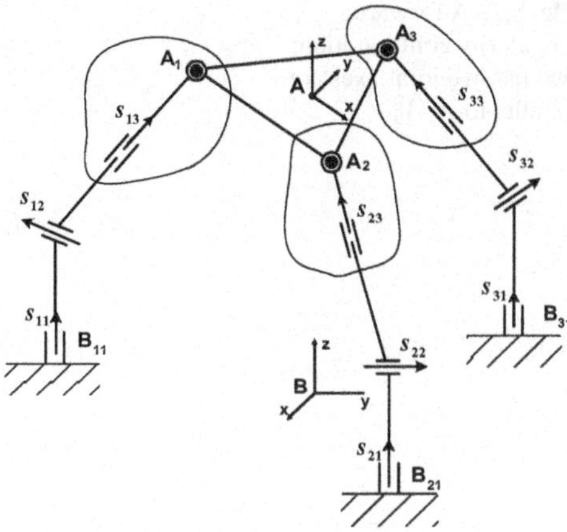

Fig. 8.5 A forward singularity configuration: passive joint axes pass through leg-end points

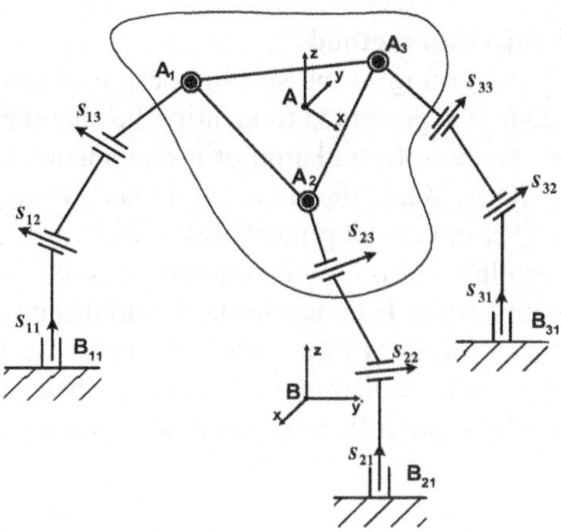

Fig. 8.6 A forward singularity configuration: one passive joint axis falls into plane $A_1A_2A_3$

Example 2: Figure 8.6 shows a specific configuration of a three-legged parallel robot, in which the passive joint axis $\hat{s}_{23}$ falls in the plane formed by points A_1, A_2, and A_3. As a result, $u_{23} \perp p_{12}$ and $u_{23} \perp p_{23}$, and hence $\Delta_p = 0$. A forward singularity occurs in this specific configuration.

- A forward singularity will occur if any two passive-joint axes, $\hat{s}_{i3}$ and $\hat{s}_{j3}$ $(i, j = 1, 2, 3; i \neq j)$, are simultaneously parallel with the vector defined by points A_i and A_j, p_{ij} (or p_{ji}). It follows that both u_{i3} and u_{j3} will be perpendicular to vector p_{ij} (or p_{ji}). At least one of the row vectors in J_p vanishes and hence, Δ_p=0.

Example 3: Figure 8.7 shows a specific configuration of a three-legged parallel robot, in which two passive joint axes $\hat{s}_{23}$ and $\hat{s}_{33}$ are simultaneously parallel with vector $\overrightarrow{A_2A_3}$, i.e., $w_{23} \parallel w_{33} \parallel p_{23}$. As a result, $u_{23} \perp p_{23}$ and $u_{33} \perp p_{23}$, and the determinant of J_p vanishes. A forward singularity occurs in this specific configuration.

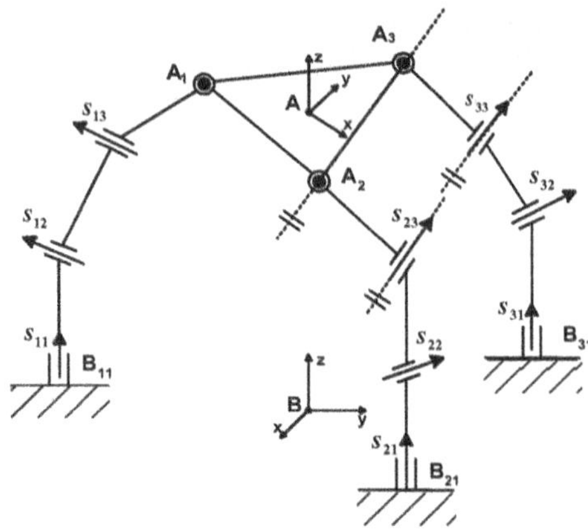

Fig. 8.7 A forward singularity configuration: two passive joint axes are parallel to A_2A_3

Projection method

According to the specific design of the 3RPRS PKM with DMA, a projection method is presented to identify the singularity conditions. With the simple geometric method, formulation of instantaneous kinematics and Jacobian matrices can be avoided. When the six active joints are rigidly locked, the PKM will be reduced to a 3RS closed-loop mechanism, as shown in Fig. 8.8. In a general case, it has no mobility and is a rigid structure. In some special configurations, however, the moving platform becomes unstable and instantaneous motions in certain directions may occur, which are exactly the cases of forward singularities. Based on the passive-joint-velocity method proposed in [185], whenever the moving platform has instantaneous motions, the following equation will always hold:

$$J_p\dot{q}_p = 0, \tag{8.35}$$

where $\dot{q}_p = (\dot{q}_{d1}, \dot{q}_{d2}, \dot{q}_{d3})$ represents the velocity of the three passive revolute joints, and matrix J_p is given by

$$J_p = \begin{bmatrix} p_{12}^T(r_1 \times \omega_{d1}) & -p_{12}^T(r_2 \times \omega_{d2}) & 0 \\ 0 & p_{23}^T(r_2 \times \omega_{d2}) & -p_{23}^T(r_3 \times \omega_{d3}) \\ -p_{31}^T(r_1 \times \omega_{d1}) & 0 & p_{31}^T(r_3 \times \omega_{d3}) \end{bmatrix},$$

where $p_{ij} = p_j - p_i, r_i = \overrightarrow{\|D_iP_i\|}$ and $\omega_{di} = (0, 0, 1)$ represents the unit directional vector of the passive revolute joint $D_i (i, j = 1, 2, 3, i \neq j)$.

It can be depicted from Eq. (8.35) that the forward singularity will occur if and only if matrix J_p is singular so that $\dot{q}_p$ has nontrivial solutions. Hence, the complete and sufficient condition for the forward singularity of the 3RPRS PKM can be written as

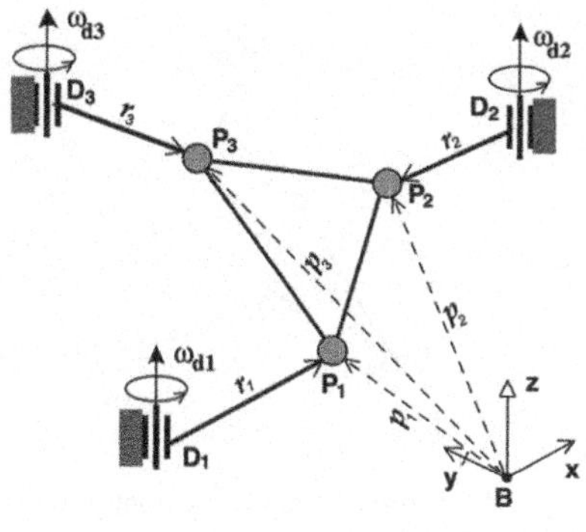

Fig. 8.8 Reduced 3RS parallel mechanism

$$\Delta_{J_p} = p_{12}^T(r_1 \times \omega_{d1}) p_{23}^T(r_2 \times \omega_{d2}) p_{31}^T(r_3 \times \omega_{d3}) \\ - p_{12}^T(r_2 \times \omega_{d2}) p_{23}^T(r_3 \times \omega_{d3}) p_{31}^T(r_1 \times \omega_{d1}) = 0. \quad (8.36)$$

Although Eq. (8.36) gives the mathematical expression of the forward singularity condition, it is still not geometrically clear. For this specific PKM with DMA, we can assume $p_i = \left(p_{ix}, p_{iy}, p_{iz}\right)$ and $r_i = \left(r_{ix}, r_{iy}, 0\right)$. Through symbolic computation, we realized that Δ_{J_p} can also be written as

$$\Delta_{J_p} = \Delta_V \Delta_H, \quad (8.37)$$

where $\Delta_V = (p_{12} \times p_{23})$. Δ_H represents the determinant of matrix H, which is given by

$$\Delta_H = \begin{bmatrix} r_{1y} & -r_{1x} & r_{1x}p_{1y} - r_{1y}p_{1x} \\ r_{2y} & -r_{2x} & r_{2x}p_{2y} - r_{2y}p_{2x} \\ r_{3y} & -r_{3x} & r_{3x}p_{3y} - r_{3y}p_{3x} \end{bmatrix} \in \Re^{3\times 3}.$$

It can also be symbolically verified that Δ_{J_p} is identical to the determinant of the 6×6 forward kinematics Jacobian matrix of the 3RPRS PKM with DMA. With Eq. (8.37), we can now geometrically conduct the forward singularity analysis. Since Δ_V represents the projection of the normal of the triangular plate $P_1P_2P_3$ onto the vertical direction ω_{di}, Δ_V will take a value of zero if the triangular plate is perpendicular to the base plane, and in turn $\Delta_{J_p} = 0$. Therefore, the PKM with DMA will be always a forward singularity configuration when the titling angle of the moving platform reaches 90°.

On the other hand, if the triangular plate is not perpendicular to the base plane such that $\Delta_V \neq 0$, the forward singularity condition will only be dependent on the value of Δ_H, i.e., $\Delta_{J_p} = 0$ if and only if $\Delta_H = 0$. It can be further verified that Δ_H is identical to the determinant of the forward kinematics Jacobian of the 3RRR planar PKM [191] obtained by projecting the 3RPRS PKM with DMA onto a horizontal plane. In other words, if the triangular plate is not perpendicular to the base plane, the forward singularity conditions can be derived from the projected 3RRR planar

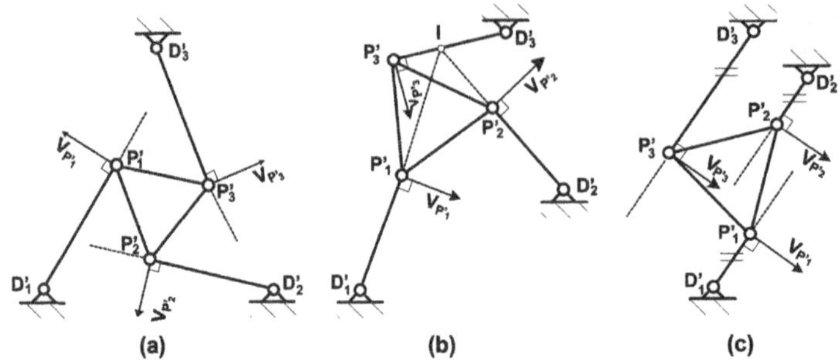

Fig. 8.9 Forward singularity configurations

PKM, which has been well studied in [191] based on the concept of instantaneous centers. Here, we will only summarize the conclusions.

By projecting the 3RS closed-loop mechanism (Fig. 8.8) onto a horizontal plane, e.g., the base plane, a planar 3RR closed-loop mechanism (without mobility in a general configuration) can be obtained, as indicated in Fig. 8.9a. Based on the method proposed in [191], it can be concluded that the forward singularity will occur if and only if the three projected links $D_1^{'}P_1^{'}$, $D_2^{'}P_2^{'}$ and $D_3^{'}P_3^{'}$ either intersect at a point Fig. 8.9b or are parallel to each other Fig. 8.9c. If the three links intersect at a point I as shown in Fig. 8.9b, the projected moving platform can produce an infinitesimal rotation about the instantaneous center I. If the three links are parallel to each other as shown in Fig. 8.9c, the projected moving platform can produce an infinitesimal translation relative to the base such that the instantaneous center lies at infinity. Otherwise, the moving platform and the base have an infinite number of instantaneous centers of zero velocity so that the moving platform is stationary with respect to the base.

Now, let us revert to the original 3RPRS PKM (Fig. 8.3). The revolute joint axis $\hat{s}_{di}$ and the center of spherical joint P_i of leg i can define a vertical plane. Hence, three such vertical planes can be formed. The forward singularity occurs if and only if the three vertical planes intersect at the same vertical line or they are parallel to each other. Moreover, if the moving platform is perpendicular to the base plane, it is always a forward singularity configuration.

8.4.2 *Inverse Singularity*

The inverse singularity configuration refers to a specific robot configuration in which the moving platform loses one or more degrees of freedom, instantaneously. In other words, the active joint velocities cannot be determined from the given moving platform velocity at an inverse singularity configuration.

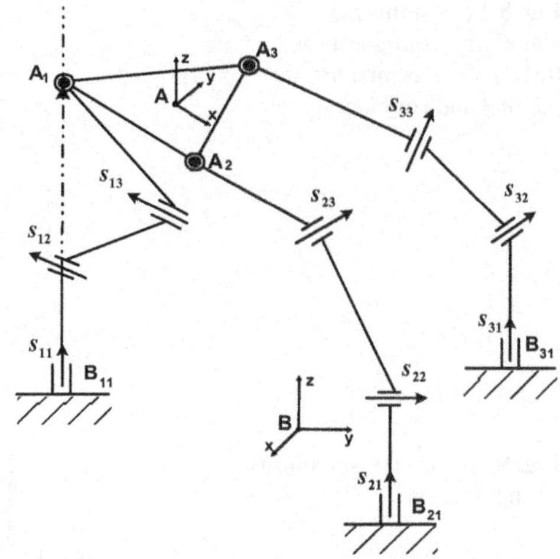

Fig. 8.10 An inverse singularity configuration: one joint axis passes through its corresponding leg-end point

Passive-joint-velocity based method

For the 3RRRS PKM, the inverse singularity occurs when any matrix $J_i \in \Re^{3\times 3}$ $(i = 1, 2, 3)$ is singular. If J_i is singular, neither the passive- nor the active-joint velocities can be determined from the mobile platform velocity. Obviously, if the determinant of matrix J_i, denoted by Δ_i, is equal to zero, then an inverse singularity will occur. Since $J_i = [u_{i1} \;\; u_{i2} \;\; u_{i3}]$, the inverse singularity can also be numerically identified when the active- and passive-joint angles are known. In fact, the inverse singularity analysis of the three-legged parallel robot is equivalent to that of three 3-DOF serial-type robot manipulators. Therefore, we can readily list two useful geometric conditions for an inverse singularity:

- An inverse singularity will occur if any revolute joint axis $\hat{s}_{ij}$ $(i, j = 1, 2, 3)$ passes through its corresponding leg-end point A_i. When this happens, the contribution of the ijth joint velocity to the point velocity of A_i vanishes, i.e., $u_{ij} = 0$. As a result, $\Delta_i = 0$.
  Example 4: Figure 8.10 shows a specific configuration of a three-legged parallel robot, in which the joint axis $\hat{s}_{11}$ passes through its corresponding leg-end point A_1. As a result, $u_{11} = 0$ and $\Delta_1 = 0$. An inverse singularity occurs in this specific configuration.
- An inverse singularity will occur if three revolute joint axes $\hat{s}_{ij} (j = 1, 2, 3)$ of leg i are parallel and coplanar. When this happens, the contributions of the three joint velocities to the point velocity of A_i, i.e., u_{i1}, u_{i2}, and u_{i3}, will be linearly dependent. As a result, $\Delta_i = 0$.
  Example 5: Figure 8.11 shows a specific configuration of a three-legged parallel robot, in which the three revolute joint axes in leg 1, i.e., $\hat{s}_{11}$, $\hat{s}_{12}$, and $\hat{s}_{13}$, are parallel and coplanar. Therefore, u_{i1}, u_{i2}, and u_{i3} are linearly dependent such that $\Delta_i = 0$. An inverse singularity occurs in this particular configuration.

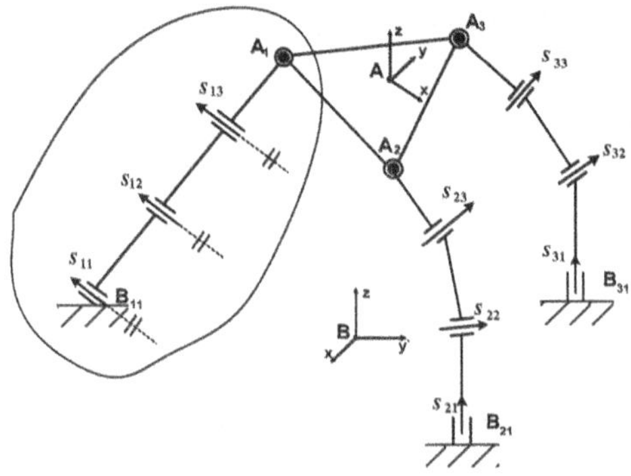

Fig. 8.11 An inverse singularity configuration: three joint axes in a leg are parallel and coplanar

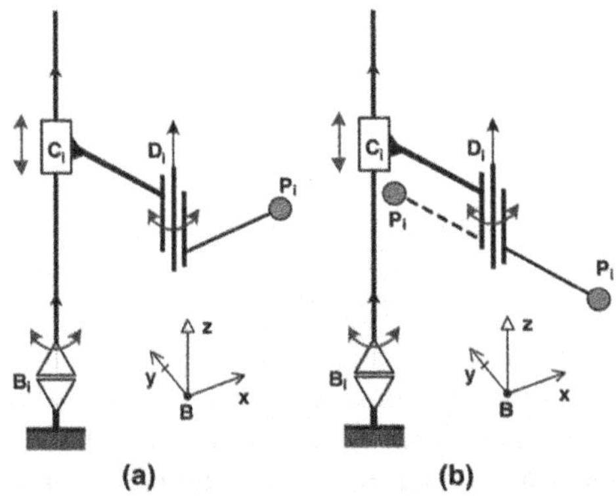

Fig. 8.12 Inverse singularity configurations

Projection method

For the 3RPRS PKM with DMA, each leg is connected to the moving platform through a spherical joint so that the point velocity of P_i can be uniquely determined with the given moving platform velocity. Hence, the inverse singularity analysis can be decoupled for each of the legs. Since each leg is a SCARA-like 3-DOF RPR type positional arm with an operational point P_i $(i = 1, 2, 3)$, at the leg-end, its inverse singularity conditions are identical with those of a SCARA-type robot, i.e., the inverse singularity occurs if and only if the two links (lines) C_iD_i and D_iP_i in leg i are co-planar, as shown in Fig. 8.12. Geometrically, this singularity condition corresponds to the fully stretched or folded configurations. The complete 3RPRS PKM with DMA will suffer from the inverse singularity if any leg is in an inverse singularity configuration.

8.4.3 Combined Singularity

The combined singularity configuration refers to a specific robot configuration in which the moving platform simultaneously gains and loses one or more degrees of freedom. Hence, the combined singularity occurs if and only if both forward and inverse singularities simultaneously occur. For the 3RRRS PKM, the example 1 is

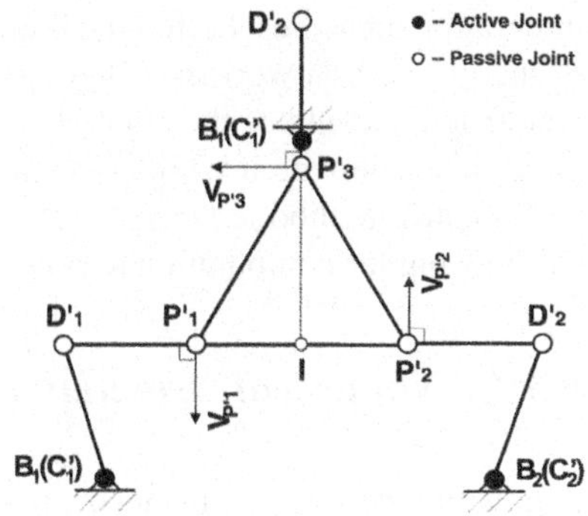

Fig. 8.13 Combined singularity configuration

actually the case of combined singularity because both matrices J_p and J_1 are simultaneously singular. For the 3RPRS PKM, the combined singularity configuration is shown in Fig. 8.13. It is noted that the combined singularity is subject to strict conditions and can be avoided by proper dimension design of the mechanisms [192], [191]. For the specific PKM being developed, there are no combined singularities in the workspace.

8.5 Workspace Analysis

The size of the workspace is one of the important indicators to measure the performance of a PKM. For a 6-DOF PKM, its complete 6-D workspace is difficult to present, the major subject of this section is to determine the subsets of the workspace. There are basically two approaches to determine the subsets of the workspace of a 6-DOF PKM, i.e., the constant orientation workspace and the constant position workspace. The constant-orientation workspace, i.e., the reachable workspace of a PKM with a constant moving platform orientation, has been well addressed by Gosselin [195], in which the constant-orientation workspace is given by the intersection of individual ones generated by each of the three legs. However, the constant position workspace, i.e., the orientation workspace of a PKM with a constant moving platform position, has not been well studied.

To investigate the constant position workspace for a 6-DOF PKM, a numerical approach is proposed [194]. With Euler angles, tilt-and-torsion angles, and exponential coordinates parameterization methods, the constant orientation workspace of a 6-DOF PKM can be represented by a rectangular parallel piped, a solid cylinder, and a solid sphere, respectively. Different finite partition schemes have been proposed for the three solid geometric entities. They possess the same essential features, i.e., they are all equi-volumetric, parametric, proportional, and exact partitions of their corresponding geometric entities. It is noticed that, as a result of parameterizations, integration measures have to be introduced when computing the orientation workspace volume from its parametric domains. Consequently, the orientation workspace volume can be numerically computed as a weighted volume sum of the equi-volumetric elements in which the weightages are the element associated integration measures. Following this approach, various global performance measures of the orientation workspace can also be readily computed. A comparison study is conducted to evaluate the effectiveness of the three parameterization methods on the numerical orientation workspace analysis. It has shown that the matrix exponential coordinates method is more effective because it has no formulation singularity and exhibits higher computation accuracy.

8.5.1 Numerical Orientation Workspace Analysis

Among the various parameterization methods for spatial rigid-body rotations, Euler angles, tilt-and-torsion angles, and exponential coordinates have proven to be appropriate for finite partitions.

8.5.1.1 Euler Angles Representation

In robotics, the Euler angles parameterization method is the most popular convention to describe the spatial rigid body rotations. Depending on the sequences of rotations and the coordinate frames used to describe the rotations, a number of Euler-angle sets can be defined. Here, we only take the widely used Z-Y-Z Euler angles representation as an example to describe its applications in numerical orientation workspace analysis. In fact, the other sets of Euler angles have almost the same features in these aspects. The rotation matrix resulting from the Z-Y-Z Euler angles is given by:

$$R_{ZYZ} = \begin{bmatrix} c\alpha c\beta c\gamma - s\alpha s\gamma & -c\alpha c\beta s\gamma - s\alpha c\gamma & c\alpha s\beta \\ s\alpha c\beta c\gamma + c\alpha s\gamma & -s\alpha c\beta s\gamma + c\alpha c\gamma & s\alpha s\beta \\ -s\beta c\gamma & s\beta s\gamma & c\beta \end{bmatrix}, \tag{8.38}$$

where 'c' and 's' represent the '$cosine$' and '$sine$' functions respectively. α, β and γ are the respective Z-Y-Z Euler angles. If we limit the range of the Euler angles such that $0 \le \alpha < 2\pi$, $0 \le \beta < \pi$, and $0 \le \gamma < 2\pi$, the entire rigid body rotation group $SO(3)$ can be visualized as a rectangular parallel piped as shown in Fig. 8.14. Thus, a one-to-one mapping can be established between the Euler angles triplets and the rigid body rotations.

As a result of Euler angles parameterization, integration measures need to be introduced when computing the volume of the orientation workspace of rigid body rotations from its parametric domain. The integration measure is defined as the absolute value of the determinant of the Jacobian matrix which relates the rate of change of the parameters to describe the rotation to the angular velocity of the rigid body. For Euler angles parameterization, it has been realized that the integration measure is given by $\sin\beta$. Hence, the volume of the entire rigid body rotation group is given by:

$$\int_{SO(3)} dV_R = \int_0^{2\pi}\int_0^{\pi}\int_0^{2\pi} \sin\beta d\alpha d\beta d\gamma = 8\pi^2. \tag{8.39}$$

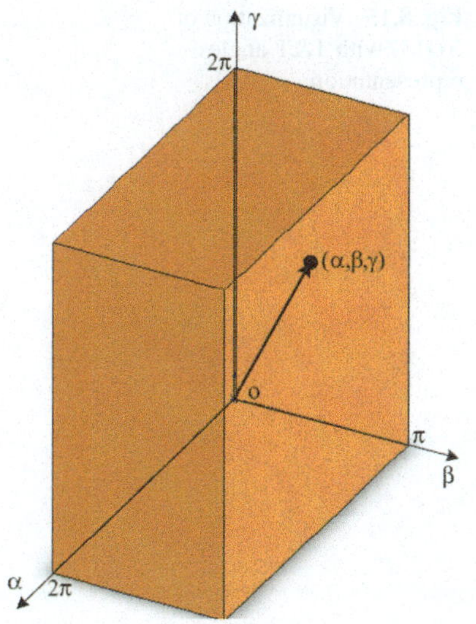

Fig. 8.14 Visualization of $SO(3)$ with Euler angles representation

Consequently, integration or convolution of a rotation-dependant function $f(R)$ over a set of rotations $S \in SO(3)$ in the Euler angles domain is given by

$$\int_{R \in S} f(R) dV_R = \int\int\int_{\alpha,\beta,\gamma \in Q_e} f(R[\alpha, \beta, \gamma]) \sin\beta d\alpha d\beta d\gamma, \tag{8.40}$$

where Q_e denotes the parameter space of the ZYZ Euler angles, i.e., a subset of the rectangular parallel piped.

8.5.1.2 Tilt-and-Torsion Angles Representation

A very intuitive parameterization method of rigid body rotations is that of Tilt-and-Torsion angles, i.e., T&T angles. The T&T angles are defined in two stages: a tilt and a torsion. This does not, however, mean that only two angles define the T&T angles but simply that the axis of tilt is variable and is defined by another angle. It is actually a three-angle orientation representation and can be considered as a modified Euler angles set.

It has been verified that the T&T angles (θ, ϕ, σ) are equivalent to the ZYZ Euler angles $(\phi, \theta, \sigma - \phi)$. Therefore, the rotation matrix resulting from the T&T angles (θ, ϕ, σ), $R_{T\&T}$ is given by:

$$\begin{bmatrix} c\phi c\theta c(\sigma-\phi) - s\phi s(\sigma-\phi) & -c\phi c\theta s(\sigma-\phi) - s\phi c(\sigma-\phi) & c\phi s\theta \\ s\phi c\theta c(\sigma-\phi) + c\phi s(\sigma-\phi) & -s\phi c\theta s(\sigma-\phi) + c\phi c(\sigma-\phi) & s\phi s\theta \\ -s\theta c(\sigma-\phi) & s\theta s(\sigma-\phi) & c\theta \end{bmatrix}. \tag{8.41}$$

Fig. 8.15 Visualization of $SO(3)$ with T&T angles representation

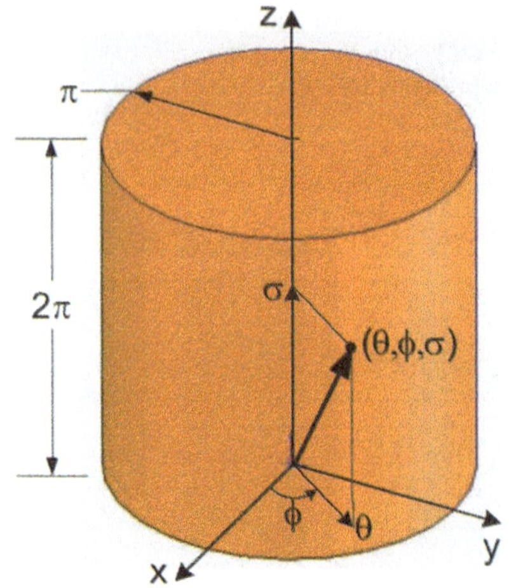

It has been addressed that the entire rigid-body rotation group $SO(3)$ can be visualized as a solid cylinder when the cylindrical coordinates are employed to represent the T&T angles, as shown in Fig. 8.15. Furthermore if we limit the ranges of the T&T angles to $0 \le \theta < \pi$, $0 \le \phi < 2\pi$, and $0 \le \sigma < 2\pi$, a one-to-one mapping can be established between T&T angles and rigid body rotations.

The integration measures also need to be introduced when computing the volume of the orientation workspace of rigid body rotations from the T&T angles domain. It can be verified that if Cartesian coordinates are used to represent the T&T angles, the integration measure will be the same as that of the Euler angles representation, which is given by $\sin\theta$. In this case, the volume of the entire rigid body rotation group is given by:

$$\int_{SO(3)} dV_R = \int_0^{2\pi}\int_0^{\pi}\int_0^{2\pi} \sin\theta d\phi d\theta d\sigma. \tag{8.42}$$

However, since the T&T angles are represented in cylindrical coordinates, the term $d\theta d\phi d\sigma$ in Eq. (8.42) is not a volume element. Then, we need to modify the term into its equivalent form as: $\frac{\theta d\phi d\theta d\sigma}{\theta}$, where $d\theta d\phi d\sigma$ is the standard differential volume element in cylindrical coordinates. Equation (8.42) can be rewritten as:

$$\int_{SO(3)} dV_R = \int_0^{2\pi}\int_0^{\pi}\int_0^{2\pi} \frac{\sin\theta}{\theta}\theta d\phi d\theta d\sigma = 8\pi^2. \tag{8.43}$$

Equation (8.43) also implies that the equivalent integration measure for the cylindrical coordinates representation of T&T angles is $\frac{\sin\theta}{\theta}$. The integration or convolution of a rotation-dependant function $f(R)$ over a set of rotations $S \in SO(3)$ in the T&T angles domain is given by

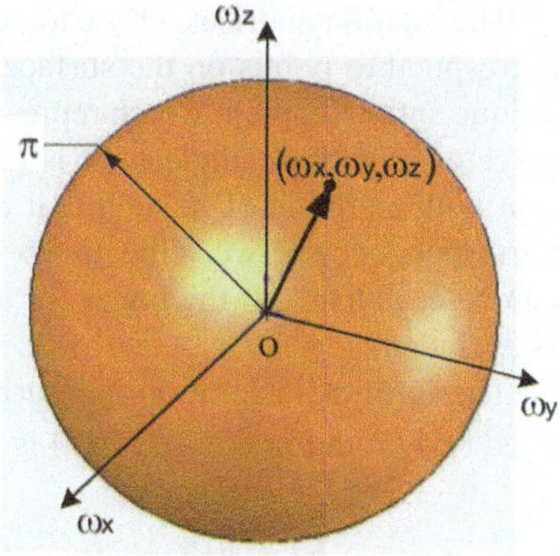

Fig. 8.16 Visualization of $SO(3)$ with exponential coordinates representation

$$\int_{R\in S} f(R)dV_R = \int\int\int_{\theta,\phi,\sigma\in Q_t} f(R[\theta,\phi,\sigma])\frac{\sin\theta}{\theta}\theta d\phi d\theta d\sigma, \qquad (8.44)$$

where Q_t denotes the parameter space of the T&T angles (with cylindrical coordinates representation), i.e., a subset of the solid cylinder.

8.5.1.3 Exponential Coordinates Representation

The exponential coordinates parameterization method is a mathematically elegant representation of rigid body rotations. It has been realized that the rigid body rotation group, i.e., the Special Orthogonal Group $SO(3)$, when embodied in $\Re^3$, can be geometrically visualized as a solid sphere of radius π, centered at the origin with the antipodal points as shown in Fig. 8.16. A point $w = (w_x, w_y, w_z)$ within the sphere represents a rotation by an angle $\|w\|$ $\left(\|w\|^2 = w_x^2 + w_y^2 + w_z^2\right)$ about the line passing through the origin and w (directed from the origin to w). In other words, the rigid body rotations can be parametrically represented by the coordinates of the points within the sphere.

Based on differential geometry and group theory, the rotation matrix associated with a point coordinate $w = (w_x, w_y, w_z)$ can be obtained through a matrix exponential mapping defined as follows:

$$e^{\hat{w}} = I + \frac{\sin\|w\|}{\|w\|}\hat{w} + \frac{1-\cos\|w\|}{\|w\|^2}\hat{w}^2, \qquad (8.45)$$

where $\hat{w} = \begin{bmatrix} 0 & -w_z & w_y \\ w_z & 0 & -w_x \\ -w_y & w_x & 0 \end{bmatrix}$ is a skew-symmetric matrix associated with w and $\|w\|^2 = w_x^2 + w_y^2 + w_z^2 \leq \pi^2$.

The rotation matrices whose traces are equal to -1 have a rotation angle of π, and correspond to points on the surface of the solid sphere. Each of these points has a unique antipodal point which represents the same rotation because the rotation about a vector w with an angle of π is equivalent to a rotation about its opposite vector $-w$ with an angle of π such that $e^{\hat{w}} = e^{\widehat{-w}}$ if $\|w\| = \pi$. Therefore, a one-to-one correspondence between the matrix exponential coordinates and rigid body rotations can be established if the lower (or upper) hemispherical surface of the solid sphere is excluded.

The form of integration measure is $\frac{2(1-\cos\|w\|)}{\|w\|^2}$. Hence, the volume of the entire rigid body rotation group $SO(3)$ is given by:

$$\int_{SO(3)} dV_R = \int_{D^3} \frac{2\,(1-\cos\|w\|)}{\|w\|^2} dw_x dw_y dw_z = 8\pi^2. \tag{8.46}$$

where D^3 represents the solid sphere of radius π.

The integration or convolution of a rotation-dependant function $f(R)$ over a set of rotations $S \in SO(3)$ in the exponential coordinates domain can be given by

$$\int_{R\in S} f(R)dV_R = \int_{w\in Q_m} f(R\,[w]) \frac{2\,(1-\cos\|w\|)}{\|w\|^2} dw_x dw_y dw_z, \tag{8.47}$$

where Q_m denotes the parameter space of the exponential coordinates, i.e., a subset of the solid sphere.

8.5.2 Finite Partition Schemes

8.5.2.1 Isotropic Partition of the Rectangular Parallel Piped

As discussed in last section, the rigid body rotation group, when embodied in $\Re^3$ through ZYZ Euler angles $(\alpha - \beta - \gamma)$, can be visualized as a rectangular parallel piped of dimensions, where $0 \leq \alpha < 2\pi$, $0 \leq \beta < \pi$, and $0 \leq \gamma < 2\pi$. A straightforward way to isotropically partition this rectangular parallel piped is to divide the ranges of α, β and γ angles into $2n$, n, and $2n$ segments with the same interval $\frac{\pi}{n}$, respectively. In this way, the entire rectangular parallel piped can be partitioned into $4n^3$ identical cubes, as shown in Fig. 8.17. Consequently, each of the cubic elements has the same volume of $V_e = \frac{\pi^3}{n^3}$. For indexing purpose, a triplet (i, j, k) is assigned to each of the elements, where $i = 1, 2, ..., 2n$ is the index for α angles starting from the origin, $j = 1, 2, ..., n$ is the index for β angles starting from the origin, and $k = 1, 2, ..., 2n$ is the index for γ angles also starting from the origin. When a triplet index (i, j, k) is specified for a cubic element, we can readily compute the coordinates of its center point as:

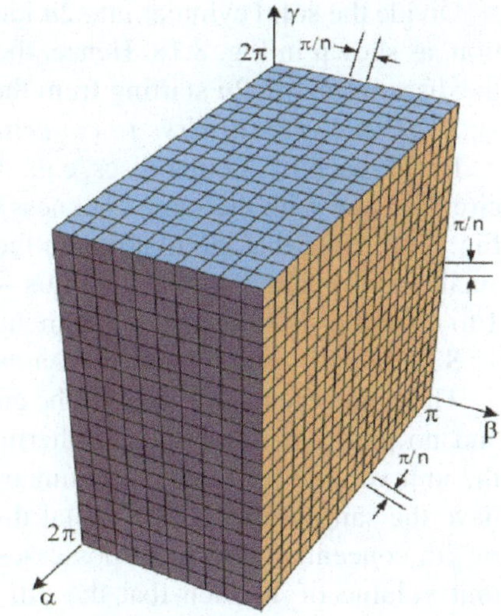

Fig. 8.17 Isotropic Partition of a rectangular parallel piped

$$
\begin{aligned}
\alpha_{ijk} &= \frac{(2i-1)\,\pi}{2n},\\
\beta_{ijk} &= \frac{(2j-1)\,\pi}{n},\\
\gamma_{ijk} &= \frac{(2k-1)\,\pi}{2n}.
\end{aligned}
\tag{8.48}
$$

When the number of division n is sufficiently large, the local property of each element can be represented by its center point, termed the feature point. According to Eq. (8.39), the orientation workspace for a set of rotations $S \in SO(3)$ can be numerically computed as:

$$\int_{R\in S} dV_R \approx V_e \sum \sin\beta_{ijk}. \tag{8.49}$$

Consequently, Eq. (8.40) can be written as:

$$\int_{R\in S} f(R)\, dV_R \approx V_e \sum f\left(R\left[\alpha_{ijk}, \beta_{ijk}, \gamma_{ijk}\right]\right) \sin\beta_{ijk}, \tag{8.50}$$

where $(\alpha_{ijk}, \beta_{ijk}, \gamma_{ijk}) \in Q_e$.

8.5.2.2 Equi-volumetric Partition of the Solid Cylinder

Based on the geometric features of the solid cylinder of radius π and height 2π, an equi-volumetric partition scheme is achieved in a number of steps as follows:

STEP 1: *Solid cylinder to circular discs*

Divide the solid cylinder into $2n$ identical thin circular discs along vertical direction, as shown in Fig. 8.18. Hence, the height of each thin circular disc is $\frac{\pi}{n}$. Label the discs from 1 to $2n$ starting from the bottom, i.e., $i = 1, 2, ..., 2n$.

STEP 2: *Circular disc to concentric circular bands*

Divide each of circular discs, e.g., the $i^{th}(i = 1, 2, ..., 2n)$ disc, into n concentric circular band with the same thickness ($\sigma\theta = \frac{\pi}{n}$) along radial direction, as shown in Fig. 8.18. Note that the first one in the center, i.e., the center element, is actually a small and thin circular disc of radius $\frac{\pi}{n}$ and height $\frac{\pi}{n}$. Label the circular bands from 1 to n starting from the center element, i.e., $j = 1, 2, ..., n$.

STEP 3: *Circular band to elements*

The center element in each of the circular discs is considered as the basic element and no partition is needed. Furthermore, its volume, i.e., $V_t = \frac{\pi^4}{n^3}$, is defined as the unit volume for the equi-volumetric partition scheme. Since the circular bands have the same thickness (in radial direction), it is not surprising that the volume of jth concentric ring band possesses $m_j = j^2 - (j-1)^2 = 2j - 1$ times of the unit volume of V_t such that the jth circular band can be partitioned into $m_j = 2j - 1$ elements with identical volume of V_t, $j = 2, 3, ..., n$. It follows that we can circumferentially divide the jth circular band into $2j - 1$ equi-volumetric elements using the same spacing angle $\frac{2\pi}{2j-1}$, as shown in Fig. 8.18. For the jth circular band, label the partitioned $2j - 1$ elements using index k such that $k = 1, 2, ..., 2j - 1$, starting from the x-axis counterclockwise. In this way, each circular disc can be equi-volumetrically divided into $\sum_{j=1}^{n}(2j-1) = n^2$ elements. Altogether, the entire solid cylinder can be partitioned into $2n^3$ elements with identical volume of V_t.

Similar to the isotropic partition of the rectangular parallel piped, when a triplet index (i, j, k) is specified for an element, the coordinates of the feature points can be readily computed. When the number of division n is sufficiently large, the local property of each element can be represented by its feature point. According to Eq. (8.43), the orientation workspace for a set of rotations $S \in SO(3)$ can be numerically computed as:

$$\int_{R\in S} dV_R \approx V_t \sum \frac{\sin\theta_{ijk}}{\theta_{ijk}}. \tag{8.51}$$

Consequently, Eq. (8.39) can be written as:

$$\int_{R\in S} f(R)\, dV_R \approx V_t \sum \frac{f\left(R\left[\theta_{ijk}, \phi_{ijk}, \sigma_{ijk}\right]\right)\sin\theta_{ijk}}{\theta_{ijk}}, \tag{8.52}$$

where $(\theta_{ijk}, \phi_{ijk}, \sigma_{ijk}) \in Q_t$.

8.5.2.3 Equi-volumetric Partition of the Solid Sphere

For numerical orientation workspace analysis using matrix exponential coordinates, a novel equi-volumetric partition method has been proposed for the solid sphere of

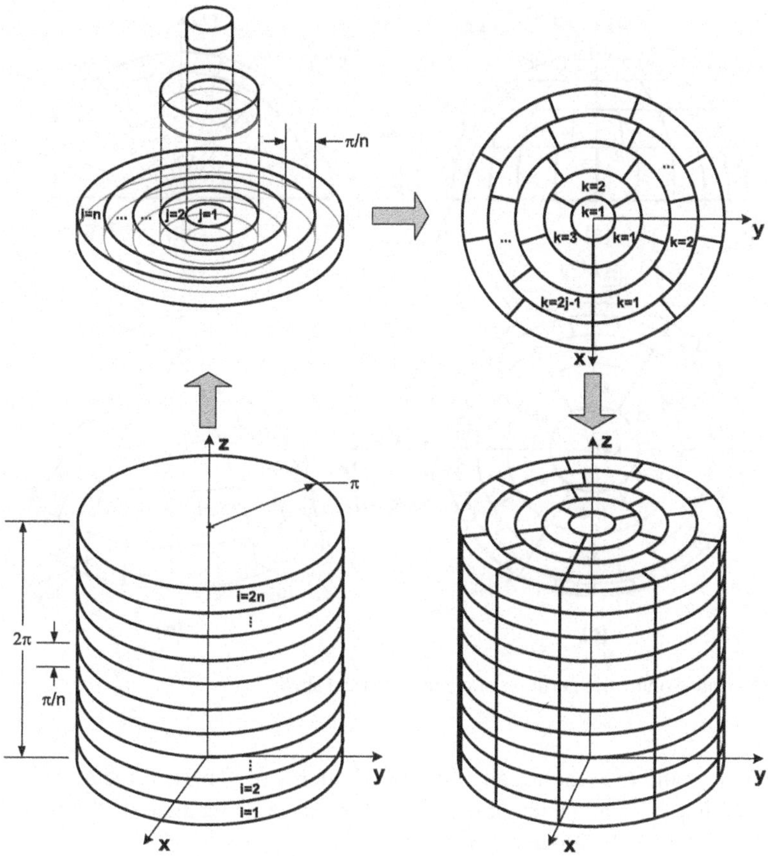

Fig. 8.18 Equi-volumetric partition of a solid cylinder

radius π. Schematically, this partition only deals with one solid hemisphere because of the geometric symmetry. The solid hemisphere is first divided into n concentric spherical shells with the same thickness. The first shell, i.e., the core, is considered as the basic element. It is actually a small solid hemisphere of radius $\frac{\pi}{n}$. Its volume $V_m = \frac{2\pi^4}{3n^3}$ is defined as the unit volume for the equi-volumetric partition scheme. Based on the geometry of the concentric spherical shell, we first latitudinally remove one circular cap element with one unit volume from the top of the shell, as shown in Fig. 8.19. Subsequently, we latitudinally partition the remaining portion of the shell into consecutive circular bands according to the volume allocations of the bands. Finally, we longitudinally partition each of the circular bands into equi-volumetric elements (identical to the volume of the core element) using the same spacing angle. Using such an equi-volumetrical partition scheme, the solid hemisphere of radius π can be divided into n^3 elements of equal volume. There is one core element of a solid hemisphere in the center, $n - 1$ cap elements of spherical disk shells along the

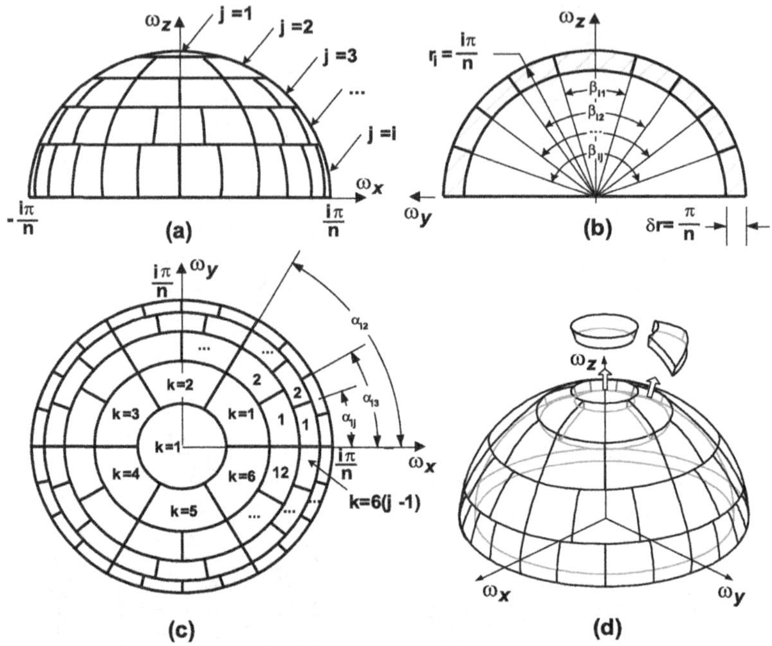

Fig. 8.19 Equi-volumetric partition of a hemispherical shell

pole axis w_z, and the rest are all common elements of spherical rectangular shells. With the other solid hemisphere partitioned in the same manner, the complete solid sphere will have $2n^3$ elements in total.

Similar to the other two partition methods, when the number of division n is sufficiently large, the local property of each element can be represented by its feature point. According to Eq. (8.46), the orientation workspace for a set of rotations $S \in SO(3)$ can be numerically computed as:

$$\int_{SO(3)} dV_R \approx 2V_m \sum \frac{\left(1 - \cos \left\| w_{ijk} \right\|\right)}{\left\| w_{ijk} \right\|^2}. \tag{8.53}$$

Consequently, Eq. (8.47) can be rewritten as:

$$\int_{R \in S} f(R)\, dV_R \approx 2V_m \sum \frac{f\left(R\left[w_{ijk}\right]\right)\left(1 - \cos \left\| w_{ijk} \right\|\right)}{\left\| w_{ijk} \right\|^2}, \tag{8.54}$$

where $w_{ijk} \in Q_m$.

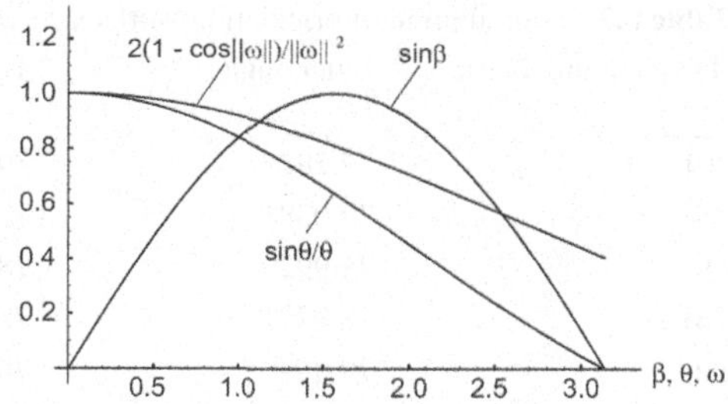

Fig. 8.20 Integration measure distributions

8.5.2.4 Comparison Study

In the previous part, three parameterization methods as well as their corresponding partition schemes have been introduced for numerical orientation workspace analysis. To have a clearer understanding about these three approaches, a comparison study in terms of integration measures, element convergency, and computation accuracy, is discussed.

Integration Measure Distribution

For the Euler angles parameterization, the integration measure is given by $\sin\beta$ $(0 \le \beta < \pi)$. For the T&T angles parameterization with cylindrical coordinates representation, the equivalent integration measure is given by $\frac{\sin\theta}{\theta}(0 \le \theta < \pi)$. Finally, the integration measure for the exponential coordinates parameterization method is given by $\frac{2(1-\cos\|w\|)}{\|w\|^2}$ $(0 \le \|w\| \le \pi)$. The distributions of these integration measures along with β, θ and $\|w\|$ are plotted in Fig. 8.20. It is clear that the integration measure for the Euler angles parameterization has singular points when $\beta = 0$ and β approaches π; the integration measure for the T&T angles has singular points when β approaches to π; whereas the exponential coordinates parameterization has no singularity in the entire parametric domain. Furthermore, with regard to the linearity level of the integration measures, the exponential coordinates parameterization is the best and the Euler angles parameterization is the worst. It also implies that the finite partition of $SO(3)$ in Euler angles parametric domain may induce the largest computation errors, whereas the exponential coordinates parameterization can give the most accurate results.

Element Convergency

In the Euler angles analysis approach, an isotropic partition scheme is proposed for the rectangular parallel piped, which is realized in three orthogonal directions. The rectangular parallel piped is divided into $4n^3$ identical cubic elements. In the T&T angles analysis approach, an equi-volumetric partition scheme is proposed for the solid cylinder, which is realized in axial, radial, and circumferential directions. The solid cylinder is divided into $2n^3$ finite elements with equal volume. In the exponential

Table 8.2 Computation of orientation workspace volume

Division number n	Euler angles	T&T angles	Exponential coordinates
10	79.2825	79.2925	79.2082
20	79.0381	79.0386	79.0196
30	78.9929	78.9930	78.9847
40	78.9771	78.9772	78.9725
50	78.9698	78.9698	78.9669
100	78.9601	78.9601	78.9593

Table 8.3 Computation errors

Division number n	Euler angles	T&T angles	Exponential coordinates
10	−0.3256	−0.3336	−0.2513
20	−0.0812	−0.0817	−0.0628
30	−0.0361	−0.0362	−0.0279
40	−0.0203	−0.0203	−0.0157
50	−0.0130	−0.0130	−0.0100
100	−0.0032	−0.0032	−0.0025

coordinates analysis approach, an equi-volumetric partition scheme is proposed for the solid sphere, which is realized in radial, latitudinal, and longitudinal directions. The solid sphere of radius π is divided into $2n^3$ finite elements with equal volume.

Regardless of the partition schemes, the 3D of each element are inversely proportional to the division number n, while its geometric volume is inversely proportional to n^3. The elements in n-division and $(n+1)$-division have the same volume ratio of $\frac{(n+1)^3}{n^3}$, which is approximately equal to unity when n is sufficiently large. Hence, it can be concluded that the elements in each of the partition schemes are all uniformly convergent when the division number n is large enough.

Computation Accuracy

A computation example is employed to compare the computation accuracy of these three numerical approaches (especially when n is not sufficiently large). In this example, the orientation workspace volume of the entire rigid-body rotation group is numerically computed with different division number n through three partition schemes, separately. It is noted the theoretical value of the orientation workspace volume, i.e., the volume of $SO(3)$, is $8\pi^2 (= 78.9568\text{rad}^2)$, regardless of the parameterization methods. The computation results are listed in Table 8.2. Based on the results listed in Table 8.2, we can further calculate the computation errors which is defined as the differences between the theoretical orientation workspace volume (which is always $8\pi^2$) and its corresponding computed values, as shown in Table 8.3. Note that with the same division number n, the number of elements in the Euler angles partition approach is twice of the other two approaches. Hence, the results shown in Table 8.3 are more generous to the Euler angles approach. It can be readily depicted that the exponential coordinates method has the highest computation accuracy, while the Euler angles approach and T&T angles approach are more or less equivalent. However, when the division number n increases, the differences among the three approaches become insignificant.

Chapter 9
Kinematic Calibration for Modular Parallel Robots

Abstract This chapter focuses on the kinematic calibration for three-legged modular parallel robots. Based on the closed-loop structure of parallel robots, the calibration procedure is normally divided into two steps, i.e., self calibration followed by base and tool calibration. Using the local frame representation of the Product-of-Exponentials (local POE) formula and mathematical tools from differential geometry and group theory, the self-calibration and base-tool calibration models are formulated. An iterative least-squares algorithm is employed to identify the error parameters. To demonstrate the accuracy, effectiveness, and robustness of these two approaches, two computation examples on a 6-DOF (3$\underline{RR}$RS) modular parallel robot are provided.

9.1 Introduction

One of the main concerns in the modular parallel robot system is the positioning accuracy of the robot end-effector. A set of robot modules is joined together to form a complete parallel robot assembly. Factors like machining tolerance, compliance, misalignment of the connected modules, and wear of the connecting mechanisms will affect the positioning accuracy of the robot. As a result, the assembly errors of a modular robot are usually larger than those of a robot having fixed configurations. Hence, identifying the critical kinematic parameters to improve the positioning accuracy of the robot end-effector becomes a very important issue for modular reconfigurable parallel robots. Based on the closed-loop structure of the parallel robot, the calibration procedure is normally divided into two steps, i.e., self-calibration of the closed-loop mechanisms and calibration of the base as well as the end-effector. The purpose of self-calibration is to calibrate the closed-loop mechanisms by using the built-in sensors in the passive joints. Once the parallel robot is self-calibrated, the subsequent end-effector calibration can be easily performed. The base-tool calibration,

G. Yang and I.-M. Chen, *Modular Robots: Theory and Practice*,
Research on Intelligent Manufacturing,
https://doi.org/10.1007/978-981-16-5007-9_9

on the other hand, is to improve the absolute positioning accuracy of the end-effector by using external measuring equipment. In this step, only the kinematic errors in the fixed transformations from the robot world frame to the robot base frame and from the mobile platform frame to the end-effector frame need to be identified.

Because of the importance of the self-calibration step, past research efforts on calibration of parallel robots have been concentrated on the self-calibration techniques [201–215]. A representative work in this approach was presented by Zhuang [216]. In this work, a self-calibration method was proposed for the conventional six-legged Stewart platform through the installation of redundant sensors in several passive joints and constructing measurement residues with the measured values and the computed values of these readable passive joint angles. When these passive joint angles were recorded at a sufficient number of measurement configurations, the actual kinematic parameters can be estimated by minimizing the measurement residues. Since this model was based on the linearization of the kinematic constraint equations, it converged rapidly. Wampler, Hollerbach, and Arai [221] presented a unified formulation for the self-calibration of both serial-link robots and robotic mechanisms having kinematically closed loops by using the implicit loop method. In this method, the kinematic errors were allocated to the joints such that the loop equations were exactly satisfied, simplifying the treatment of multi-loop mechanisms. Inrascu and Park [222] developed a unified geometric framework for the calibration of kinematic chains containing closed loops. Both joint encoder readings and end-effector pose measurements can be uniformly included in this framework. As a result, the kinematic calibration was cast as a nonlinear constrained optimization problem.

There is only a handful of works on the calibration of the three-legged manipulators [222, 223]. Notash and Podhorodeski [223] presented a methodology allowing kinematic calibration of the three-legged parallel manipulator based on the minimization of the leg-end distance error. The work employed the Lavenberg-Marquardt nonlinear least-squares algorithm to identify kinematic parameters of the manipulator.

In this chapter, following the leg-end approach, we develop a general and effective linear calibration algorithm for modular parallel robots based on the local frame representation of the Product-of-Exponentials (POE) formula. The POE representation method describes the joint axes as twists based on line geometry. It is, therefore, uniform in modeling manipulators with both revolute and prismatic joints. The kinematic parameters in the POE model vary smoothly with changes in joint axes so that the model can cope with certain formulation singularity problems that can not be handled by using other kinematic parameterization methods. Significantly, the POE formula has a very explicit differential structure such that it can be easily differentiated with respect to any of its kinematic parameters [224–226]. Since a modular reconfigurable parallel robot has unlimited configurations, it is necessary for the calibration algorithm to be as generic to robot configurations. For these reasons, the POE formula would be the most appropriate modeling tool for the modular parallel robot calibration. A simulation example of calibrating a three-leg $\underline{\text{RR}}$RS modular parallel robot shows that the robot kinematics can be fully calibrated within two to three iterations.

On the other hand, the base and tool calibration, aiming at calibrating the base and tool frames of a parallel robot after the self-calibration, also plays an important role in enhancing the accuracy of a parallel robot. The base and tool calibration of a parallel robot is similar to that of a serial type robot, and there is much work addressing such a calibration issue. Zhuang [227] et al. proposed a linear solution based on the quaternion algebra that allowed the simultaneous computation of these two kinematic transformations. This approach is non-iterative and fast. However, this linear method has the formulation singularity problem such that some arrangements of the robot world frame and the robot base frame have to be avoided. Another inconvenience of this approach is that the geometric parameters are estimated in a two-stage process so that the estimation errors from the first stage will propagate to the second stage. It will be always the case when noise and measurement errors exist. Other representative works related to the base and tool calibration can be found in Park and Martin [228], Dornaika and Horaud [229], and in [230–240].

Different from the linear solution approach, we cast the base and tool calibration problem into an iterative parameter identification process. The reason behind this is that although the linear (closed-form) solution approach, in principle, is computationally efficient, it becomes ineffective in the presence of noise and measurement errors. Using the mathematical tools from differential geometry and group theory, a simple and explicit base and tool calibration algorithm is formulated in this chapter. The basic idea of this algorithm is that the kinematic errors in a fixed kinematic transformation can be represented by a twist, i.e., an element of $se(3)$. Because a twist also has a 6-dimensional vector representation, the formulation is significantly simplified. Besides, it does not suffer from any formulation singularity problem and is generous enough to allow very large kinematic errors. The simulation examples show that it is also computational efficient as all the preset kinematic errors can be fully recovered within three to four iterations.

9.2 Self-calibration for Three-legged Modular Parallel Robots

In order to develop a self-calibration model for the three-legged modular parallel robots, the forward kinematics analysis is necessary, which has been briefly introduced in Chap. 8. This section mainly addresses the formulation of the self-calibration algorithm for a class of three-legged modular parallel robots.

9.2.1 *Self-calibration Model Based on Leg-end Distance Errors*

Due to the closed-loop structure of a parallel robot, the forward kinematic transformations of its legs are coupled together through the spherical joints and the unique mobile platform. The overall kinematic errors of a parallel robot are contributed by the kinematic errors in each of the legs and those in the mobile platform in a coupled manner. Traditionally, the self-calibration model of a parallel robot is formulated through constructing a measurement residue with measured values and computed values of the readable passive joints. Although the calibration idea is simple, the model formulation is complicated. Besides, during the parameter identification process, the passive joint displacements need to be updated at each of the iterations, which lowers the computational efficiency.

Based on the local POE formula, a simple self-calibration model is formulated for a class of three-legged parallel robots as shown in Fig. 9.1. The identification objective function is defined as the leg-end distance errors because they are very sensitive to the variations of the kinematic parameters so as to speed up the identification process. In order to simplify the calibration model, we assume that (1) the kinematic errors

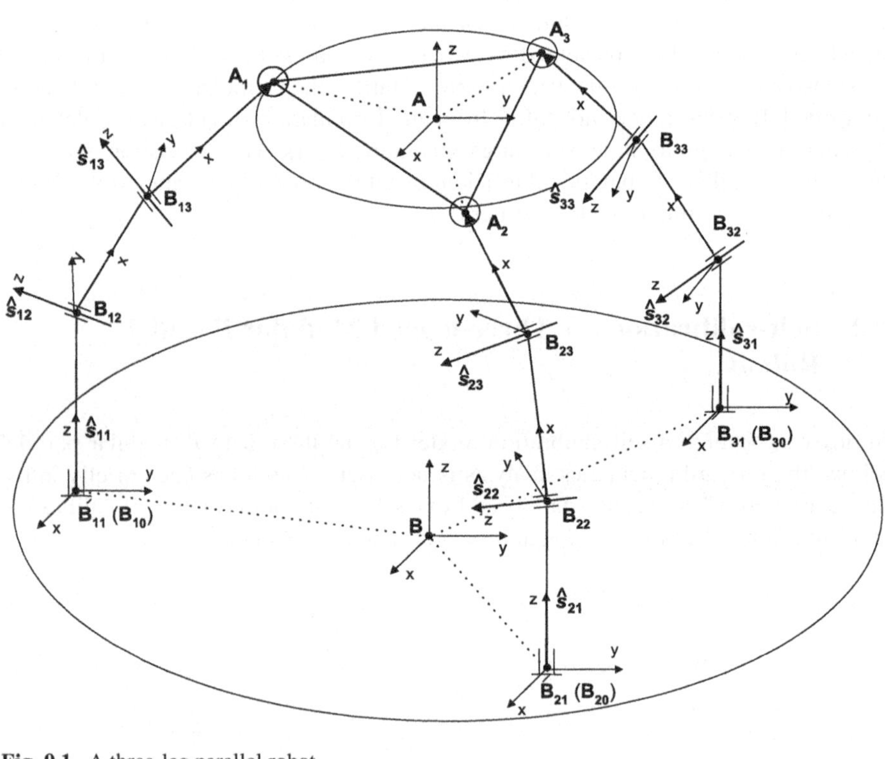

Fig. 9.1 A three-leg parallel robot

in a dyad exist only in the relative initial pose, (2) the joint twist coordinate and the joint angle offset in a dyad have no kinematic error and remain in their nominal values. Since all the readable passive joints can be considered as the active joints, they will always take their measured values during the identification process and the forward displacement analysis at each iteration becomes unnecessary. Note that in this formulation, we also assume that the 3-DOF spherical joint is kinematically perfect such that its three axes intersect at one point.

Now let us first consider the kinematic errors of an individual leg. Due to the kinematic errors in the leg assembly, the actual leg-end position will be different from its nominal value. Based on the sensor-based solution approach presented in Chap. 8, the forward kinematic transformation of leg i can also be given by:

$$\begin{bmatrix} p_i \\ 1 \end{bmatrix} = T_{B,i1}(0) e^{\hat{s}_{i1} q_{i1}} T_{i1,i2}(0) e^{\hat{s}_{i2} q_{i2}} T_{i2,i3}(0) e^{\hat{s}_{i3} q_{i3}} \begin{bmatrix} p_i' \\ 1 \end{bmatrix}, \tag{9.1}$$

where p_i' is the position vector of point A_i with respect to local frame $i3$ ($i = 1, 2, 3$). Note that the homogeneous coordinate representation is employed in Eq. (9.1). Once the position vector p_i is computed, we can determine the pose of the mobile platform [241].

According to the definition of matrix logarithm defined on $SE(3)$, there exists at least a $\hat{t} \in se(3)$ for a given $T \in SE(3)$, such that $e^{\hat{t}} = T$. Hence, for the initial pose $T_{i(j-1),ij}(0)$, it is sufficient to let $e^{\hat{t}_{ij}} = T_{i(j-1),ij}(0)$ (with $e^{\hat{t}_{i1}} = T_{B,i1}(0)$), where $\hat{t}_{ij} \in se(3)$ ($i, j = 1, 2, 3$). Equation (9.1) can be rewritten as

$$\begin{bmatrix} p_i \\ 1 \end{bmatrix} = e^{\hat{t}_{i1}} e^{\hat{s}_{i1} q_{i1}} e^{\hat{t}_{i2}} e^{\hat{s}_{i2} q_{i2}} e^{\hat{t}_{i3}} e^{\hat{s}_{i3} q_{i3}} \begin{bmatrix} p_i' \\ 1 \end{bmatrix}. \tag{9.2}$$

For each of the legs, we assume that the kinematic errors occur only in the initial pose $T_{i(j-1),ij}(0)$ (hence in $\hat{t}_{ij}$) and the position vector p_i'. Let the kinematic errors in $\hat{t}_{ij}$ be expressed in the local frame $i(j-1)$, denoted by $\delta\hat{t}_{ij}$. Since $\hat{t}_{ij} \in se(3)$, $\delta\hat{t}_{ij}$ also belongs to $se(3)$. Geometrically, $\delta e^{\hat{t}_{ij}} = \delta\hat{t}_{ij} e^{\hat{t}_{ij}}$. Linearizing Eq. (9.1) with respect to $\hat{t}_{ij}$ and p_i', we have

$$\begin{aligned} \begin{bmatrix} \delta p_i \\ 0 \end{bmatrix} &= \delta\hat{t}_{i1} e^{\hat{t}_{i1}} e^{\hat{s}_{i1} q_{i1}} e^{\hat{t}_{i2}} e^{\hat{s}_{i2} q_{i2}} e^{\hat{t}_{i3}} e^{\hat{s}_{i3} q_{i3}} \begin{bmatrix} p_i' \\ 1 \end{bmatrix} \\ &+ e^{\hat{t}_{i1}} e^{\hat{s}_{i1} q_{i1}} \delta\hat{t}_{i2} e^{\hat{t}_{i2}} e^{\hat{s}_{i2} q_{i2}} e^{\hat{t}_{i3}} e^{\hat{s}_{i3} q_{i3}} \begin{bmatrix} p_i' \\ 1 \end{bmatrix} \\ &+ e^{\hat{t}_{i1}} e^{\hat{s}_{i1} q_{i1}} e^{\hat{t}_{i2}} e^{\hat{s}_{i2} q_{i2}} \delta\hat{t}_{i3} e^{\hat{t}_{i3}} e^{\hat{s}_{i3} q_{i3}} \begin{bmatrix} p_i' \\ 1 \end{bmatrix} \\ &+ e^{\hat{t}_{i1}} e^{\hat{s}_{i1} q_{i1}} e^{\hat{t}_{i2}} e^{\hat{s}_{i2} q_{i2}} e^{\hat{t}_{i3}} e^{\hat{s}_{i3} q_{i3}} \begin{bmatrix} \delta p_i' \\ 0 \end{bmatrix}, \end{aligned} \tag{9.3}$$

where $\delta\hat{t}_{ij} \in se(3)$ is the kinematic errors in $\hat{t}_{ij}$ expressed in module frame $i(j-1)$ and $\delta p_i^{'} \in \Re^{3\times 1}$ is the kinematic error of the position vector $p_i^{'}$ with respect to frame $i3$. Based on the fact that $e^{\hat{t}_{ij}} = T_{i(j-1),ij}(0)$, Eq. (9.3) can also be simplified as

$$\begin{bmatrix} \delta p_i \\ 0 \end{bmatrix} = \delta\hat{t}_{i1} \begin{bmatrix} p_{B,i}^{'} \\ 1 \end{bmatrix} + T_{B,i1}\delta\hat{t}_{i2} \begin{bmatrix} p_{i1,i}^{'} \\ 1 \end{bmatrix} + T_{B,i2}\delta\hat{t}_{i3} \begin{bmatrix} p_{i2,i}^{'} \\ 1 \end{bmatrix} + T_{B,i3} \begin{bmatrix} \delta p_i^{'} \\ 0 \end{bmatrix}, \quad (9.4)$$

where $T_{B,ij}$ $(j = 1, 2, 3)$ represents the forward kinematic transformation from frame B to frame ij and $p_{ij,i}^{'} \in \Re^{3\times 1}$ (with $p_{i0,i}^{'} = p_{B,i}^{'}$) represents the position vector of point A_i with respect to frame ij.

Equation (9.4) is actually a differential equation. It describes the gross kinematic error of the leg-end position vector p_i resulting from the kinematic errors in the initial pose $T_{i(j-1),ij}(0)$ and the position vector $p_i^{'}$. However, Eq. (9.4) appears to have a nonlinear form, which is undesirable in robot calibration. With some modifications, Eq. (9.4) can be converted into a clear linear equation as described as follows.

Let $\delta\hat{t}$ be an element of $se(3)$ such that $\delta\hat{t} = \begin{bmatrix} \delta\hat{w} & \delta v \\ 0 & 0 \end{bmatrix}$ and $p \in \Re^{3\times 1}$ be a position vector. We have

$$\delta\hat{t} \begin{bmatrix} p \\ 1 \end{bmatrix} = \begin{bmatrix} \delta\hat{\omega} & \delta v \\ 0 & 0 \end{bmatrix} \begin{bmatrix} p \\ 1 \end{bmatrix} = \begin{bmatrix} \delta v + \delta\hat{\omega} p \\ 0 \end{bmatrix} = \begin{bmatrix} \delta v - \hat{p}\delta\omega \\ 0 \end{bmatrix} = \begin{bmatrix} I_{3\times 3} & -\hat{p} \\ 0 & 0 \end{bmatrix} \begin{bmatrix} \delta v \\ \delta\omega \end{bmatrix}. \quad (9.5)$$

In Eq. (9.5), the matrix $\left[I_{3\times 3} - \hat{p} \right] \in \Re^{3\times 6}$ can be considered as the transition matrix related to the position vector p. We term such a matrix as the positional transition matrix and denote it by T_p. Therefore, Eq. (9.5) can be rewritten as

$$\delta\hat{t} \begin{bmatrix} p \\ 1 \end{bmatrix} = \begin{bmatrix} T_p \\ 0 \end{bmatrix} \delta t, \quad (9.6)$$

where $\delta t = (\delta v, \delta w)^T \in \Re^{6\times 1}$ is a 6-dimensional vector representation of $\delta\hat{t}$.

Substituting Eq. (9.6) into Eq. (9.4), we have

$$\begin{bmatrix} \delta p_i \\ 0 \end{bmatrix} = \begin{bmatrix} T_{p_{B,i}^{'}} \\ 0 \end{bmatrix} \delta t_{i1} + T_{B,i1} \begin{bmatrix} T_{p_{i1,i}^{'}} \\ 0 \end{bmatrix} \delta t_{i2} + T_{B,i2} \begin{bmatrix} T_{p_{i2,i}^{'}} \\ 0 \end{bmatrix} \delta t_{i3} + T_{B,i3} \begin{bmatrix} \delta p_i^{'} \\ 0 \end{bmatrix}. \quad (9.7)$$

Since $T_{B,ij} = \begin{bmatrix} R_{B,ij} & p_{B,ij} \\ 0 & 1 \end{bmatrix}$, in which $R_{B,ij}$ and $p_{B,ij}$ $(i, j = 1, 2, 3)$ represent the orientation and position of frame ij with respect to the base frame B, respectively. Equation (9.7) can also be further simplified as

$$\delta p_i = T_{p_{B,i}^{'}} \delta t_{i1} + R_{B,i1} T_{p_{i1,i}^{'}} \delta t_{i2} + R_{B,i2} T_{p_{i2,i}^{'}} \delta t_{i3} + R_{B,i3} \delta p_i^{'} = J_i \delta t_i, \quad (9.8)$$

where

$$J_i = \left[T_{p'_{B,i}} \; R_{B,i1}T_{p'_{i1,i}} \; R_{B,i2}T_{p'_{i2,i}} \; R_{B,i3}\right] \in \Re^{3\times 21},$$
$$\delta t_i = \left[\delta t_{i1} \; \delta t_{i2} \; \delta t_{i3} \; \delta p'_i\right]^T \in \Re^{21\times 1}.$$

Apparently, Eq. (9.8) is a linear equation with respect to the kinematic errors. Based on this equation, a linear self-calibration model for the three-legged modular parallel robots can be formulated.

Now let us consider the differential change of the leg-end distance resulting from the kinematic errors. Without loss of generality, we first consider the leg-end distance between leg 1 and leg 2 denoted by a_{12} such that

$$a_{12}^2 = (p_2 - p_1)^T(p_2 - p_1). \tag{9.9}$$

Differentiating Eq. (9.9) with respect to p_1 and p_2, we have

$$\delta a_{12} = \frac{(p_2 - p_1)^T}{a_{12}}(\delta p_2 - \delta p_1) = \frac{(p_2 - p_1)^T}{\sqrt{(p_2 - p_1)^T(p_2 - p_1)}}(J_2\delta t_2 - J_1\delta t_1). \tag{9.10}$$

Equation (9.10) describes the differential change of a_{12} resulting from the kinematic errors in leg 1 and leg 2. Similarly, for the leg-end distance between leg 2 and leg 3: a_{23} and the leg-end distance between leg 3 and leg 1: a_{31}, we have

$$\delta a_{23} = \frac{(p_3 - p_2)^T}{\sqrt{(p_3 - p_2)^T(p_3 - p_2)}}(J_3\delta t_3 - J_2\delta t_2), \tag{9.11}$$

$$\delta a_{31} = \frac{(p_1 - p_3)^T}{\sqrt{(p_1 - p_3)^T(p_1 - p_3)}}(J_1\delta t_1 - J_3\delta t_3). \tag{9.12}$$

If we know the actual values of a_{12}, a_{23} and a_{31} denoted by a_{12}^a, a_{23}^a and a_{31}^a, respectively, the differential change of the leg-end distance can be given by

$$\delta a = \begin{bmatrix} \delta a_{12} \\ \delta a_{23} \\ \delta a_{31} \end{bmatrix} = \begin{bmatrix} a_{12}^a - a_{12} \\ a_{32}^a - a_{23} \\ a_{31}^a - a_{31} \end{bmatrix}, \tag{9.13}$$

where a_{12}, a_{23}, and a_{31} represent computed nominal values determined by the built-in sensor readings of the passive joint displacements. Therefore, they can be given by $\sqrt{(p_2 - p_1)^T (p_2 - p_1)}$, $\sqrt{(p_3 - p_2)^T (p_3 - p_2)}$, and $\sqrt{(p_1 - p_3)^T (p_1 - p_3)}$, respectively. Note that a_{12}, a_{23}, and a_{31} are different from their original theoretical values. Arranging Eqs. (9.10), (9.11), and (9.12) into a matrix form, a linear calibration model can be obtained:

$$\delta a = J\delta t, \tag{9.14}$$

where

$$\delta a = \begin{bmatrix} a_{12}^a - \sqrt{(p_2 - p_1)^T (p_2 - p_1)} \\ a_{23}^a - \sqrt{(p_3 - p_2)^T (p_3 - p_2)} \\ a_{31}^a - \sqrt{(p_1 - p_3)^T (p_1 - p_3)} \end{bmatrix} \in \Re^{3\times 1}, \quad \delta t = \begin{bmatrix} \delta t_1 \\ \delta t_2 \\ \delta t_3 \end{bmatrix} \in \Re^{63\times 1},$$

$$J = \begin{bmatrix} -\frac{(p_2-p_1)^T}{\sqrt{(p_2-p_1)^T(p_2-p_1)}} J_1 & \frac{(p_2-p_1)^T}{\sqrt{(p_2-p_1)^T(p_2-p_1)}} J_2 & 0 \\ 0 & -\frac{(p_3-p_2)^T}{\sqrt{(p_3-p_2)^T(p_3-p_2)}} J_2 & \frac{(p_3-p_2)^T}{\sqrt{(p_3-p_2)^T(p_3-p_2)}} J_3 \\ \frac{(p_1-p_3)^T}{\sqrt{(p_1-p_3)^T(p_1-p_3)}} J_1 & 0 & -\frac{(p_1-p_3)^T}{\sqrt{(p_1-p_3)^T(p_1-p_3)}} J_3 \end{bmatrix} \in \Re^{3\times 63}.$$

This linear calibration model is based on the assumption that all of the actual leg-end distances, i.e., a_{12}^a, a_{23}^a and a_{31}^a, are known. Such an assumption, however, is not always true. In many cases, we are unable to know the actual leg-end distances without external measuring equipment. A modified calibration model is therefore developed, in which the actual leg-end distances are also considered as the kinematic parameters to be identified. Denote the original theoretical values of the leg-end distances as a_{12}^0, a_{23}^0 and a_{31}^0, respectively. The actual leg-end distances, which are in the neighborhood of their original theoretical values, can be given by

$$\begin{aligned} a_{12}^a &= a_{12}^0 + \delta a_{12}^*, \\ a_{23}^a &= a_{23}^0 + \delta a_{23}^*, \\ a_{31}^a &= a_{31}^0 + \delta a_{31}^*, \end{aligned} \tag{9.15}$$

where δa_{12}^*, δa_{23}^* and δa_{31}^* represent the differential changes of the leg-end distances with respect to their original theoretical values, respectively. They are different from δa_{12}, δa_{23} and δa_{31} which represent the differential changes of the leg-end distances with respect to their computed nominal values. Substituting Eq. (9.15) into Eq. (9.14), we can obtain a modified linear calibration model such that

$$y = \mathcal{A}x, \tag{9.16}$$

where

$$y = \begin{bmatrix} a_{12}^0 - \sqrt{(p_2 - p_1)^T (p_2 - p_1)} \\ a_{23}^0 - \sqrt{(p_3 - p_2)^T (p_3 - p_2)} \\ a_{31}^0 - \sqrt{(p_1 - p_3)^T (p_1 - p_3)} \end{bmatrix} \in \Re^{3\times 1},$$

$$\mathcal{A} = \begin{bmatrix} J & I_{3\times 3} \end{bmatrix} \in \Re^{3\times 66}, \quad x = \begin{bmatrix} \delta t_1 & \delta t_2 & \delta t_3 & \delta a_{12}^* & \delta a_{23}^* & \delta a_{31}^* \end{bmatrix}^T \in \Re^{66\times 1}.$$

In the modified calibration model, we have altogether 66 error parameters to be identified, which reflect the kinematic errors of a three-legged modular parallel robot.

9.2.2 An Iterative Least-squares Algorithm

Based on the calibration model, i.e., Eq. (9.16), an iterative least-squares algorithm is employed for the calibration solution. To improve the calibration accuracy, we need to measure the passive joint displacements in many different robot postures. Suppose we need to take m sets of measured data. For the ith measurement, we can obtain a y_i, as well as an identification Jacobian matrix $\mathcal{A}_i$. After m measurements, we can stick y_i and $\mathcal{A}_i$ to form the following equation:

$$\tilde{Y} = \tilde{\mathcal{A}}x, \tag{9.17}$$

where

$$\tilde{Y} = \begin{bmatrix} y_1 \; y_2 \; \ldots \; y_m \end{bmatrix}^T \in \Re^{3m\times 1},$$
$$\tilde{\mathcal{A}} = column\begin{bmatrix} \mathcal{A}_1, \; \mathcal{A}_2, \; \ldots, \; \mathcal{A}_m \end{bmatrix} \in \Re^{3m\times 66}.$$

Since Eq. (9.17) consists of $3m$ linear equations with 66 variables (normally $m>$ 22), the linear least-squares algorithm is employed for the parameter identification. The least-square solution of x is given by

$$x = (\tilde{\mathcal{A}}^T\tilde{\mathcal{A}})^{-1}\tilde{\mathcal{A}}^T\tilde{Y}, \tag{9.18}$$

where $\left(\tilde{\mathcal{A}}^T\tilde{\mathcal{A}}\right)^{-1}\tilde{\mathcal{A}}^T$ is the pseudoinverse of $\tilde{\mathcal{A}}$. Due to the closed-loop structure of the parallel robot, the determinant of $\tilde{\mathcal{A}}^T\tilde{\mathcal{A}}$ is normally very small. To avoid the computational difficulty, the Singularity Value Decomposition (SVD) method can be employed to derive the pseudoinverse of $\tilde{\mathcal{A}}$.

The solution of Eq. (9.18) can be further improved through iterative substitution as shown in Fig. 9.2. Once the kinematic error parameter vector, x is identified, the initial pose $T_{i(j-1),ij}(0)$, the position vector $p_i^{'}$, and the actual leg-distances are updated by substituting x into the following equations:

$$\begin{aligned} T_{i(j-1),ij}(0)^{new} &= e^{\delta\hat{t}_{ij}}T_{i(j-),ij}(0)^{old}, \\ p_i^{'new} &= p_i^{'old} + \delta p_i^{'}, \\ a_{12}^{0\;new} &= a_{12}^{0\;old} + \delta a_{12}^{*}, \\ a_{23}^{0\;new} &= a_{23}^{0\;old} + \delta a_{23}^{*}, \\ a_{31}^{0\;new} &= a_{31}^{0\;old} + \delta a_{31}^{*}. \end{aligned} \tag{9.19}$$

The same procedures are repeated until the norm of the error vector, $\|x\|$, approaches zero and the actual leg-end distances converge to some stable values. Then the final $T_{i(j-1),ij}(0)$, $p_i^{'}$, a_{12}^0, a_{23}^0 and a_{31}^0 represent the calibrated kinematic parameters of robots, denoted by $T_{i(j-1),ij}^c(0)$, $p_i^{'c}$, a_{12}^c, a_{23}^c and a_{31}^c.

Note that the kinematic error vector, x, will no longer represent the actual kinematic errors after iterations. However, the actual kinematic errors can be extracted by comparing the calibrated kinematic parameters with their nominal values.

In order to evaluate the calibration result, we define a deviation metric, i.e., the average leg-end distance error, as

$$\delta a = \sqrt{\frac{1}{3m}\sum_{i=1}^{m}((a_{12}^c - a_{12}^{(i)})^2 + (a_{23}^c - a_{23}^{(i)})^2 + (a_{31}^c - a_{31}^{(i)})^2)}, \tag{9.20}$$

where $a_{12}^{(i)}$, $a_{23}^{(i)}$ and $a_{31}^{(i)}$ are the three leg-end distances computed according to the calibration results, respectively.

9.2.3 Computation Example

In this section, a simulation example of calibrating a 6-DOF 3RRRS parallel robot is given to demonstrate the effectiveness of the calibration algorithm. As shown in the kinematic diagram in Fig. 9.1, the nominal kinematic parameters of the 6-DOF modular parallel robot are given as follows:

$$T_{B,10} = \begin{bmatrix} \frac{1}{2} & -\frac{\sqrt{3}}{2} & 0 & -250 \\ \frac{\sqrt{3}}{2} & \frac{1}{2} & 0 & -250\sqrt{3} \\ 0 & 0 & 1 & 90 \\ 0 & 0 & 0 & 1 \end{bmatrix}, \quad T_{B,20} = \begin{bmatrix} -1 & 0 & 0 & 500 \\ 0 & -1 & 0 & 0 \\ 0 & 0 & 1 & 90 \\ 0 & 0 & 0 & 1 \end{bmatrix},$$

$$T_{B,30} = \begin{bmatrix} \frac{1}{2} & \frac{\sqrt{3}}{2} & 0 & -250 \\ -\frac{\sqrt{3}}{2} & \frac{1}{2} & 0 & 250\sqrt{3} \\ 0 & 0 & 1 & 90 \\ 0 & 0 & 0 & 1 \end{bmatrix}, \quad T_{i0,i1} = \begin{bmatrix} 0 & 0 & 1 & 90 \\ 0 & 1 & 0 & 0 \\ -1 & 0 & 0 & 0 \\ 0 & 0 & 0 & 1 \end{bmatrix},$$

$$T_{i1,i2} = \begin{bmatrix} 0 & 0 & -1 & -330 \\ 0 & -1 & 0 & 0 \\ -1 & 0 & 0 & 0 \\ 0 & 0 & 0 & 1 \end{bmatrix}, \quad T_{i2,i3} = \begin{bmatrix} 0 & 1 & 0 & 0 \\ 0 & 0 & -1 & 0 \\ -1 & 0 & 0 & 0 \\ 0 & 0 & 0 & 1 \end{bmatrix},$$

$$p_1'' = \begin{bmatrix} -205 \\ -205\sqrt{3} \\ 0 \end{bmatrix}, \quad p_2'' = \begin{bmatrix} 410 \\ 0 \\ 0 \end{bmatrix}, \quad p_3'' = \begin{bmatrix} -205 \\ 205\sqrt{3} \\ 0 \end{bmatrix}, \quad p_i'' = \begin{bmatrix} 0 \\ -330 \\ 0 \end{bmatrix},$$

$$s_{i1} = s_{i2} = s_{i3} = [0, 0, 0, 0, 0, 1]^T.$$

Here p_i'' $(i = 1, 2, 3)$ represents the nominal leg-end positions with respect to the mobile platform frame. Hence, we can compute the original theoretical values of the leg-end distances as:

$$a_{12}^0 = \sqrt{(p_2'' - p_1'')^T (p_2'' - p_1'')} = 410\sqrt{3},$$

$$a_{23}^0 = \sqrt{(p_3'' - p_2'')^T (p_3'' - p_2'')} = 410\sqrt{3}\,,$$
$$a_{31}^0 = \sqrt{(p_1'' - p_3'')^T (p_1'' - p_3'')} = 410\sqrt{3}.$$

Note that the units of the kinematic parameters are in radians and millimeters. The following procedures are employed for the simulation of the calibration algorithm.
1. Generate 50 robot poses as well as the corresponding 50 sets of nominal passive joint angles by using the numerical forward kinematic algorithm;
2. Assign errors at the kinematic parameters such as dt_{ij}, dq_{ij}, dp_i' and $da = (da_{12}, da_{23}, da_{31})^T$, where $i = 1, 2, 3$ (listed in Table 9.1);
3. Find the actual initial poses in each dyad: $T_{i(j-1),ij}^a(0) = T_{i(j-1),ij}(0)e^{d\hat{t}_{ij}}$, the actual position vector of each leg end with respect to frame $i3$: $p_i'^a = p_i' + dp_i'$ and the actual leg-end distances: $a_{12}^a = a_{12}^0 + da_{12}$, $a_{23}^a = a_{23}^0 + da_{23}$, $a_{31}^a = a_{31}^0 + da_{31}$;
4. Determine the simulated actual passive joint displacements for the 50 poses using the numerical forward kinematics algorithm;
5. Identify the kinematic errors by using the iterative calibration algorithm.

Since each of the actuators and passive joints are assumed to be a true 1-DOF joint, the condition for the assignment of errors in each of the joint twists must be satisfied such that $\|w_i + dw_i\| = 1$ and $(w_i + dw_i)^T(v_i + dv_i) = 0$, where $s_i = (v_i, w_i)^T$ and $ds_i = (dv_i, dw_i)^T$. Moreover, in the actual calibration experiment, all of the actual joint displacements, including both active and passive joints, can be directly obtained from the joints encoder readings.

The calibrated initial local frame poses as well as the kinematic errors are listed in Table 9.2. Since the preset and identified errors do not have the same physical meaning and are not one-to-one correspondence, the preset kinematic errors are not fully recovered. Note that the calibration solution is not unique and not necessarily identical to the actual robot. However, the success of the calibration simulation can be deduced from the results shown in Fig. 9.3, where the average leg-end distance error (combined for the 50 poses) is reduced from about 15 mm to nearly 0 within 3 iterations. This result shows that under the calibrated parameters description, we can directly employ the nominal joint twist coordinates and the joint displacements from both actuator and passive joint encoder readings to compute the actual kinematics of the parallel robot. In other words, the parallel robot itself is precisely calibrated.

Table 9.1 Preset kinematic errors

Parameter	Preset errors	Parameter	Preset errors
dt_{ij}	$(2, 2, 2, 0.02, 0.02, 0.02)^T$	$dq_{ij}(j \neq 3)$	0.02
ds_{ij}	$(0, 0, 0, 0, \sin(0.02), \cos(0.02) - 1)^T$	$dq_{ij}(j = 3)$	0
dp_i''	$(2, 2, 2)^T$	da	$(2, 2, 2)^T$

Table 9.2 Identified kinematic errors

Dyad	Kinematic errors	$T^{c}_{i(j-1),ij}(0)$
0-1(Leg1)	$(-7.698, 7.941, -0.160, 0.01763, 0.02284, -0.00001)^T$	$\begin{bmatrix} -0.02284 & -0.86569 & 0.50005 & -210.664 \\ 0.01763 & 0.49976 & 0.86599 & -348.699 \\ 0.99958 & 0.02859 & 0.00384 & 88.383 \\ 0 & 0 & 0 & 1 \end{bmatrix}$
1-2(Leg1)	$(1.36772, 8.42726, 3.04695, 0.00027, 0.01880, 0.02956)^T$	$\begin{bmatrix} 0.5964 & -0.6385 & 0.4864 & -117.155 \\ 0.8001 & 0.5217 & -0.2962 & 216.306 \\ -0.0647 & 0.5658 & 0.8220 & 445.614 \\ 0 & 0 & 0 & 1 \end{bmatrix}$
2-3(Leg1)	$(2.022, -0.097, -1.296, 0.05999, 0.00135, -0.00058)^T$	$\begin{bmatrix} 0.5964 & -0.6385 & 0.4864 & -117.155 \\ 0.8001 & 0.5217 & -0.2962 & 216.306 \\ -0.0647 & 0.5658 & 0.8220 & 445.614 \\ 0 & 0 & 0 & 1 \end{bmatrix}$
0-1(Leg2)	$(-2.570, -10.644, -0.171, -0.02861, -0.00384, -0.00013)^T$	$\begin{bmatrix} 0.00384 & -0.00018 & -0.99999 & 407.081 \\ 0.02860 & -0.99959 & 0.00007 & -8.100 \\ -0.99958 & 0.02860 & -0.00384 & 91.514 \\ 0 & 0 & 0 & 1 \end{bmatrix}$
1-2(Leg2)	$(1.368, 8.427, 2.581, 0.00027, 0.01880, 0.02955)^T$	$\begin{bmatrix} -0.01879 & 0.02954 & -0.99939 & -328.530 \\ -0.00001 & -0.99956 & -0.02955 & -1.304 \\ -0.99982 & -0.00054 & 0.01879 & 8.771 \\ 0 & 0 & 0 & 1 \end{bmatrix}$
2-3(Leg2)	$(2.022, -0.096, -1.301, 0.05999, 0.00127, -0.00058)^T$	$\begin{bmatrix} -0.00125 & 0.99999 & -0.00062 & 2.021 \\ 0.05995 & -0.00054 & -0.99820 & -0.057 \\ -0.99820 & -0.00124 & -0.05995 & -1.304 \\ 0 & 0 & 0 & 1 \end{bmatrix}$
0-1(Leg3)	$(10.271, 2.697, -0.086, 0.01097, -0.01899, -0.00007)^T$	$\begin{bmatrix} 0.01899 & 0.86585 & 0.49994 & -196.414 \\ 0.01097 & 0.49982 & -0.86606 & 356.794 \\ -0.99976 & 0.02194 & -0.00001 & 90.006 \\ 0 & 0 & 0 & 1 \end{bmatrix}$
1-2(Leg3)	$(1.377, 8.436, 2.776, 0.01033, 0.01891, 0.02278)^T$	$\begin{bmatrix} -0.01903 & 0.02268 & -0.99956 & -328.548 \\ 0.01011 & -0.99969 & -0.02288 & 0.888 \\ 0.99977 & -0.01055 & 0.01879 & 9.008 \\ 0 & 0 & 0 & 1 \end{bmatrix}$
2-3(Leg3)	$(2.022, -0.080, -1.316, 0.05999, 0.00106, -0.01057)^T$	$\begin{bmatrix} -0.00075 & 0.99994 & -0.01060 & 2.021 \\ 0.05996 & -0.01053 & -0.99815 & -0.052 \\ -0.99820 & -0.00138 & -0.05995 & -1.31884 \\ 0 & 0 & 0 & 1 \end{bmatrix}$
Position vector	Kinematic errors	$p_i^{\prime c}$
$p_1^{\prime}$	$(-4.000, 4.598, 0.057)^T$	$(-4.000, -325.402, 0.057)^T$
$p_2^{\prime}$	$(-4.025, 4.598, 0.058)^T$	$(-4.025, -325.402, 0.058)^T$
$p_3^{\prime}$	$(-3.994, 4.598, 0.084)^T$	$(-3.994, -325.402, 0.084)^T$
Leg-end distance	Kinematic errors	$(a^c_{12}, a^c_{23}, a^c_{31})^T$
$(1-2, 2-3, 3-1)$	$(-3.599, -3.599, -3.599)^T$	$(706.542, 706.542, 706.542)^T$

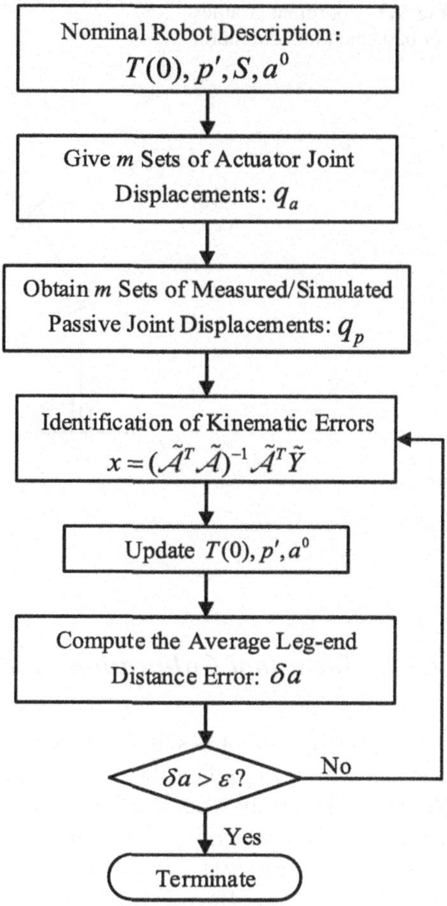

Fig. 9.2 Iterative calibration loop

9.3 Base-tool Calibration for Three-leg Modular Reconfigurable Parallel Robots

In this section, we focus on the simultaneous base and tool calibration of a self-calibrated parallel robot. After the self-calibration of a parallel robot by using the built-in sensors in the passive joints, its kinematic transformation from the robot base to the mobile platform frame can be computed with sufficient accuracy. The base and tool calibration, hence, is to identify the kinematic errors in the fixed transformations from the world frame to the robot base frame and from the mobile platform frame to the tool (end-effector) frame in order to improve the absolute positioning accuracy of the robot.

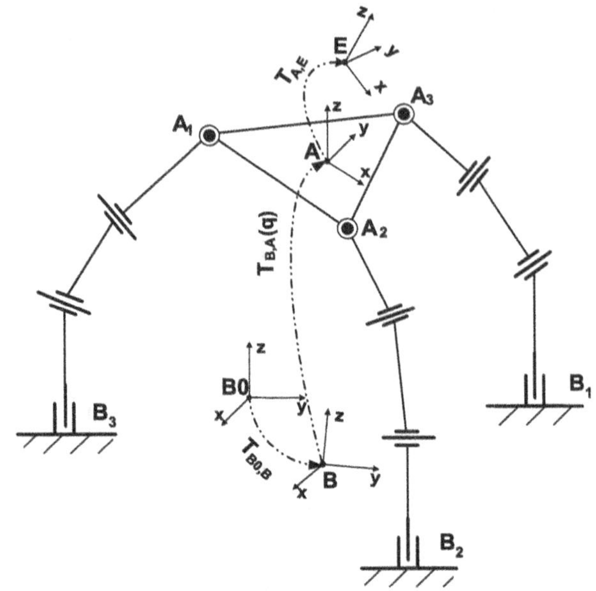

Fig. 9.3 Coordinate frames for base and tool calibration

9.3.1 Base-tool Calibration Model Based on POE Formula

Once a parallel robot is self-calibrated, as shown in Fig. 9.3, the forward kinematic transformation from the robot base frame B to the mobile platform frame A can be computed precisely when the actuator joint angles are known. Hence, the base-tool calibration is to identify the kinematic errors in the fixed transformations from the robot world frame $B0$ to the robot base frame B and from the mobile platform frame A to the tool (end-effector) frame E.

Now let us first consider the kinematic transformation from the robot base frame $B0$ to the end-effector frame E. As shown in Fig. 9.3, this forward kinematic transformation, $T_{B0,E}(q)$, can be given by

$$T_{B0,E}(q) = T_{B0,B}T_{B,A}(q)T_{A,E}, \tag{9.21}$$

where $T_{B0,B}$ and $T_{A,E}$ are the fixed kinematic transformations from the robot world frame $B0$ to the robot base frame B and from the mobile platform frame A to the tool frame E, respectively. Since $T_{B,A}(q)$ is the forward kinematic transformation of the self-calibrated parallel robot, we can assume that $T_{B,A}(q)$ is error-free and can be computed with sufficient accuracy. The kinematic errors will only come from the fixed kinematic transformations $T_{B0,B}$ and $T_{A,E}$.

According to the definition of matrix logarithm defined on $SE(3)$, there exists at least a $\hat{t} \in se(3)$ for a given $T \in SE(3)$, such that $e^{\hat{t}} = T$. Hence, it is sufficient to let $e^{\hat{t}_{B0,B}} = T_{B0,B}$ and $e^{\hat{t}_{A,E}} = T_{A,E}$, where $\hat{t}_{B0,B}, \hat{t}_{A,E} \in se(3)$. Equation (9.21) can be rewritten as:

$$T_{B0,E}(q) = e^{\hat{t}_{B0,B}} T_{B,A}(q) e^{\hat{t}_{A,E}}. \tag{9.22}$$

Now we assume that the kinematic errors occur only in $T_{B0,B}$ and $T_{A,E}$, hence in $\hat{t}_{B0,B}$ and $\hat{t}_{A,E}$. Let the kinematic errors in $\hat{t}_{B0,B}$ be expressed in the robot world frame B_0, denoted by $\delta\hat{t}_{B0,B}$, and $\hat{t}_{A,E}$ be expressed in the mobile platform frame A, denoted by $\delta\hat{t}_{A,E}$. Since both $\hat{t}_{B0,B}$ and $\hat{t}_{A,E}$ belong to $se(3)$, $\delta\hat{t}_{B0,B}$ and $\delta\hat{t}_{A,E}$ will belong to $se(3)$. Also we have $\delta e^{\hat{t}_{B0,B}} = \delta\hat{t}_{B0,B} e^{\hat{t}_{B0,B}}$ and $\delta e^{\hat{t}_{A,E}} = \delta\hat{t}_{A,E} e^{\hat{t}_{A,E}}$. Linearizing Eq. (9.21) with respect to $\hat{t}_{B0,B}$ and $\hat{t}_{A,E}$, we have

$$\delta T_{B0,E}(q) = \delta\hat{t}_{B0,B} e^{\hat{t}_{B0,B}} T_{B,A}(q) e^{\hat{t}_{A,E}} + e^{\hat{t}_{B0,B}} T_{B,A}(q) \delta\hat{t}_{A,E} e^{\hat{t}_{A,E}}. \tag{9.23}$$

Right-multiplicating both sides of Eq. (9.23) with $T_{B0,E}^{-1}(q)$, we get:

$$\delta T_{B0,E}(q) T_{B0,E}^{-1}(q) = \delta\hat{t}_{B0,B} + T_{B0,A}(q) \delta\hat{t}_{A,E} T_{B0,A}^{-1}(q), \tag{9.24}$$

where

$$\begin{aligned} T_{B0,A}(q) &= T_{B0,B} T_{B,A}(q), \\ \delta T_{B0,E}(q) T_{B0,E}^{-1}(q) &\approx T_{B0,E}^{a}(q) T_{B0,E}^{-1}(q) - I_{4\times 4}. \end{aligned}$$

Here, $T_{B0,E}^{a}(q)$ is the actual (measured) end-effector pose. Based on the definition of the matrix logarithm and the adjoint representation on $SE(3)$, we have:

$$\begin{aligned} \log\left[T_{B0,E}^{a}(q) T_{B0,E}^{-1}(q)\right] &\approx T_{B0,E}^{a}(q) T_{B0,E}^{-1}(q) - I_{4\times 4}, \\ T_{B0,A}(q) \delta\hat{t}_{A,E} T_{B0,A}^{-1}(q) &= Ad_{T_{B0,A}(q)} \delta\hat{t}_{A,E}. \end{aligned}$$

Hence, Eq. (9.24) can be rewritten as:

$$\log\left[T_{B0,E}^{a}(q) T_{B0,E}^{-1}(q)\right]^{\vee} = \delta t_{B0,B} + Ad_{T_{B0,A}(q)} \delta t_{A,E}, \tag{9.25}$$

where $\log\left[T_{B0,E}^{a}(q) T_{B0,E}^{-1}(q)\right]^{\vee} \in \Re^{6\times 1}$ is the vector representation of $\log\left[T_{B0,E}^{a}(q) T_{B0,E}^{-1}(q)\right] \in se(3)$. Similarly, $\delta t_{B0,B}$ and $\delta t_{A,E}$ are the 6-dimensional vector representations of $\delta\hat{t}_{B0,B} \in se(3)$ and $\delta\hat{t}_{A,E} \in se(3)$, respectively. Geometrically, $\log\left[T_{B0,E}^{a}(q) T_{B0,E}^{-1}(q)\right]^{\vee}$ represents the gross kinematic error of the tool frame expressed in the robot world frame. From Eq. (9.25), it is equal to the sum of the kinematic errors in $T_{B0,B}$ and $T_{A,E}$ expressed in the robot world frame, i.e., the sum of the world frame representation of $\delta t_{B0,B}$ and $\delta t_{A,E}$. Note that, in each of the 6-dimensional error vectors $\delta t_{B0,B}$ and $\delta t_{A,E}$, the first three parameters represent the position errors (δx, δy, δz), while the last three parameters represent the orientation errors ($\delta\alpha_x$, $\delta\beta_y$, $\delta\gamma_z$). A detailed geometrical description of these parameters can be obtained by using the math tools in Appendix B. With such a geometrical treatment, the formulation of the calibration model can be systematically simplified. Equation (9.25) can also be further simplified as the following linear calibration model:

$$y = Jx, \tag{9.26}$$

where

$$y = \log\left[T^a_{B0,E}(q)T^{-1}_{B0,E}(q)\right]^\vee \in \Re^{6\times 1},$$
$$J = \left[I_{6\times 6} Ad_{T_{B0,A}(q)}\right] \in \Re^{6\times 12},$$
$$x = \begin{bmatrix} \delta t_{B0,B} \\ \delta t_{A,E} \end{bmatrix} \in \Re^{12\times 1}.$$

In Eq. (9.26), we totally have 12 parameters to be identified, which can reflect kinematic errors existed in the whole system.

9.3.2 An Iterative Least-squares Algorithm

Based on the calibration model Eq. (9.26), an iterative least-squares algorithm is employed for the calibration solution. To obtain the reliable calibration result, we usually need to measure the end-effector poses in a number of different robot postures. Suppose we need to take m sets of measured data. For the ith measurement, we can obtain a $y_{[i]}$ as well as an identification Jacobian matrix $J_{[i]}$. After m measurements, we can stack $y_{[i]}$ and $J_{[i]}$ to form the following equation:

$$\tilde{y} = \tilde{J}x, \tag{9.27}$$

where

$$\tilde{y} = \begin{bmatrix} y_{[1]} \\ y_{[2]} \\ \vdots \\ y_{[m]} \end{bmatrix} \in \Re^{6m\times 1},\ \tilde{J} = \begin{bmatrix} J_{[1]} \\ J_{[2]} \\ \vdots \\ J_{[m]} \end{bmatrix} \in \Re^{6m\times 12},\ x = \begin{bmatrix} \delta t_{B0,B} \\ \delta t_{A,E} \end{bmatrix} \in \Re^{12\times 1}.$$

Since Eq. (9.27) consists of $6m$ linear equations with 12 variables (normally $m>2$), the linear least-squares algorithm is employed for the parameter identification. The least-square solution of x is given by

$$x = (\tilde{J}^T\tilde{J})^{-1}\tilde{J}^T\tilde{y}, \tag{9.28}$$

where $(\tilde{J}^T\tilde{J})^{-1}\tilde{J}^T$ is the pseudoinverse of $\tilde{J}$.

The solution of Eq. (9.28) can be further improved through iterative substitution as shown in Fig. 9.4. Once the kinematic error parameter vector, x, is identified, $T_{B0,B}$ and $T_{A,E}$ are updated by substituting x into the following equations:

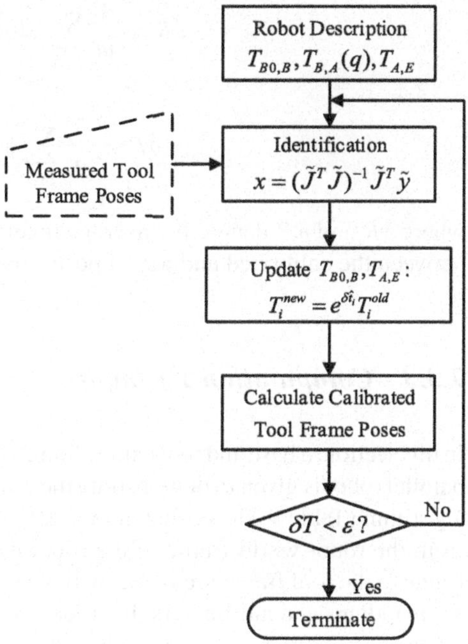

Fig. 9.4 Iterative calibration loop

$$\begin{aligned} T_{B0,B}^{new} &= e^{\delta \hat{t}_{B0,B}} T_{B0,B}^{old}, \\ T_{A,E}^{new} &= e^{\delta \hat{t}_{A,E}} T_{A,E}^{old}. \end{aligned} \tag{9.29}$$

The same procedure is repeated until the norm of the error vector, $\|x\|$, approaches zero. Then the final $T_{B0,B}$ and $T_{A,E}$ represent the calibrated kinematic transformations, denoted by $T_{B0,B}^c$ and $T_{A,E}^c$, respectively.

Note that the kinematic error vector, x, will no longer represent the actual kinematic errors after iterations. However, the actual kinematic errors can be extracted by using the matrix logarithm to compare the calibrated frame poses with their nominal counterparts such that:

$$\begin{aligned} \delta t_{B0,B} &= \log \left[T_{B0,B}^c(0) T_{B0,B}^{-1}(0) \right]^\vee, \\ \delta t_{A,E} &= \log \left[T_{A,E}^c(0) T_{A,E}^{-1}(0) \right]^\vee. \end{aligned} \tag{9.30}$$

After calibration, the forward kinematic equation becomes

$$T_{B0,E}^c(q) = T_{B0,B}^c T_{B,A}(q) T_{A,E}^c. \tag{9.31}$$

In order to evaluate the calibration result, we define the deviation metrics between the measured (actual) 'a' and calibrated 'c' end-effector (tool) frames mathematically as

$$\delta R = \frac{1}{m}\sum_{i=1}^{m} \| \log(R_{ai}^{-1} R_{ci})^{\vee} \|, \tag{9.32}$$

$$\delta P = \frac{1}{m}\sum_{i=1}^{m} \| P_{ai} - P_{ci} \|, \tag{9.33}$$

where δR and δP denote the average quantified orientation and position deviations between the calibrated and actual poses, respectively.

9.3.3 Computation Example

In this section, a base and tool calibration example for a self-calibrated 6-DOF 3RRRS parallel robot is given to demonstrate the effectiveness of the base and tool calibration algorithm (Table 9.3). As shown in Fig. 9.3, the nominal kinematic transformations from the robot world frame to the robot base frame and from the mobile platform frame to the tool frame are listed in Table 9.4. The units of the kinematic parameters are in radians and millimeters. In order to verify the robustness of the proposed algorithm, the pose of $T_{B0,B}$ is specially arranged. Such an arrangement will be a singular configuration for the closed-form method [227]. In this example, we first simulate the calibration process under an ideal experimental condition without measurement noise. To obtain the actual tool frame poses, the following procedures are employed:

Table 9.3 Preset and identified kinematic errors

	Preset errors	Identified errors
$\delta t'_{B0,B}$	$(20, 20, 20, 0.2, 0.2, 0.2)^T$	$(20, 20, 20, 0.2, 0.2, 0.2)^T$
$\delta t'_{A,E}$	$(-25, -25, -25, -0.25, -0.25, -0.25)^T$	$(-25, -25, -25, -0.25, -0.25, -0.25)^T$

Table 9.4 Nominal and calibrated $T_{B0,B}$ and $T_{A,E}$

	Nominal	Calibrated
$T_{B0,B}$	$\begin{bmatrix} -1 & 0 & 0 & 0 \\ 0 & 0 & 1 & 0 \\ 0 & 1 & 0 & 0 \\ 0 & 0 & 0 & 1 \end{bmatrix}$	$\begin{bmatrix} -0.9604 & 0.2158 & -0.1762 & 20 \\ -0.2158 & -0.1762 & 0.9604 & 20 \\ 0.1762 & 0.9604 & 0.2158 & 20 \\ 0 & 0 & 0 & 1 \end{bmatrix}$
$T_{A,E}$	$\begin{bmatrix} 1 & 0 & 0 & 0 \\ 0 & 1 & 0 & 0 \\ 0 & 0 & 1 & 100 \\ 0 & 0 & 0 & 1 \end{bmatrix}$	$\begin{bmatrix} 0.9358 & 0.2730 & -0.2115 & -46.150 \\ -0.2115 & 0.9358 & 0.2730 & 2.303 \\ 0.2730 & -0.2115 & 0.9358 & 68.847 \\ 0 & 0 & 0 & 1 \end{bmatrix}$

Table 9.5 Three given mobile platform poses $T_{B,A}(q)$

Posture	$T_{B,A}(q)$
1	$\begin{bmatrix} 0.5964 & -0.6385 & 0.4864 & -117.155 \\ 0.8001 & 0.5217 & -0.2962 & 216.306 \\ -0.0647 & 0.5658 & 0.8220 & 445.614 \\ 0 & 0 & 0 & 1 \end{bmatrix}$
2	$\begin{bmatrix} 0.6219 & -0.7604 & 0.1873 & -226.347 \\ 0.7372 & 0.6491 & 0.1875 & -57.298 \\ -0.2642 & 0.0215 & 0.9642 & 571.754 \\ 0 & 0 & 0 & 1 \end{bmatrix}$
3	$\begin{bmatrix} 0.7082 & -0.7013 & -0.0811 & 126.003 \\ 0.4699 & 0.5540 & -0.6872 & -120.154 \\ 0.5268 & 0.4486 & 0.7220 & 421.954 \\ 0 & 0 & 0 & 1 \end{bmatrix}$

1. Assign kinematic errors at $T_{B0,B}$ and $T_{A,E}$ by introducing $\delta t'_{B0,B}$ and $\delta t'_{A,E}$, respectively (listed in Table 9.3);
2. Find the actual poses of $T_{B0,B}$ and $T_{A,E}$ such that

$$T_{B0,B} = e^{\delta \tilde{t}_{B0,B}} T_{B0,B},$$
$$T_{A,E} = e^{\delta \tilde{t}_{A,E}} T_{A,E};$$

3. Calculate the actual end-effector frame poses at different robot postures

$$T^a_{B0,E}(q) = T^a_{B0,B} T_{B,A}(q) T^a_{A,E}.$$

Here, the number of measured poses is set to 3, which is the necessary condition (lower bond) for identifying the kinematic errors with perfect measurement data. The three given mobile platform poses $T_{B,A}(q)$ are listed in Table 9.5. The success of the calibration simulation can be deduced from Table 9.3, where all the preset kinematic errors are fully recovered although the preset kinematic errors are significantly large. A further re-enforcement of the calibration results is from Fig. 9.5, both the quantified orientation and position deviations, as in Eqs. (9.32) and (9.33), are driven from the initial magnitude of 160.450 mm and 0.2391 rad to values approximately zero with three to four iterations.

However, if the measurement noise exists, more measured end-effector poses will be considered. In the same example, we finally illustrate the effect of the measurement noise and robot repeatability on the calibration result. From the viewpoint of computer simulation, both the measurement noise and robot repeatability can be simulated in a uniform way by adding uniformly distributed noise on each of the actual (measured) end-effector poses. The resulting actual end-effector poses $T^{a*}_{B0,E}$ can be given by

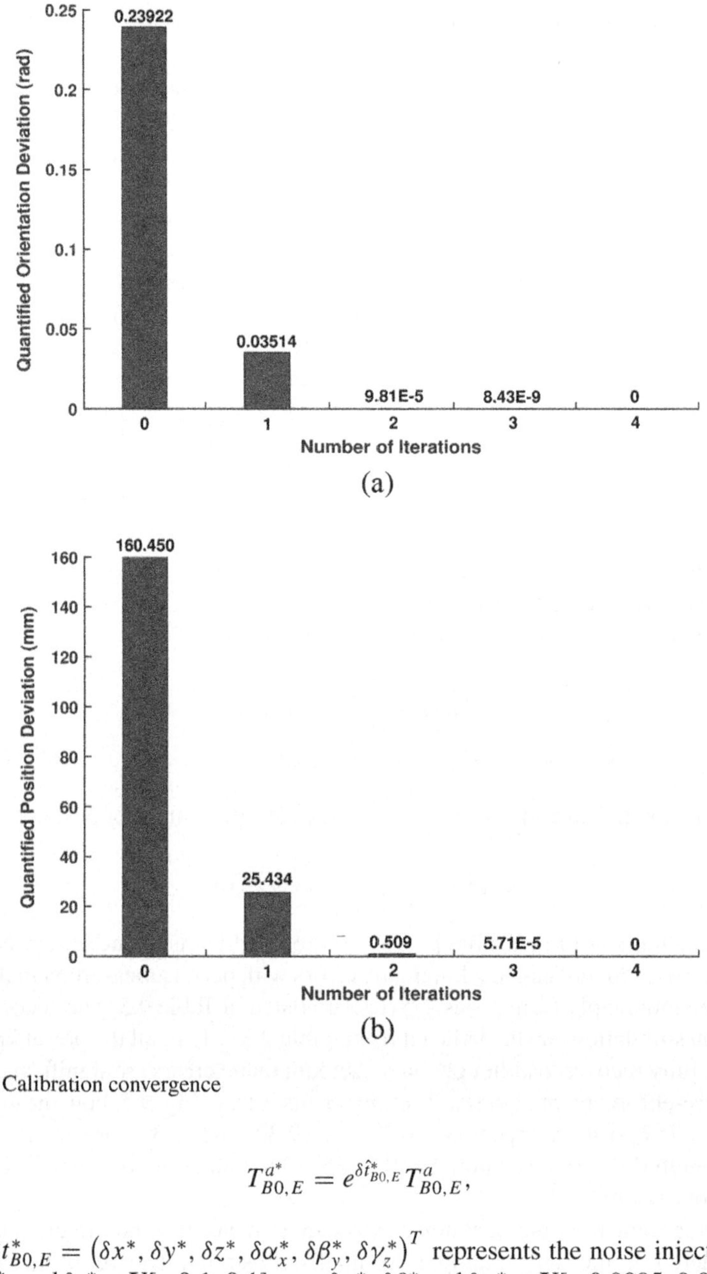

Fig. 9.5 Calibration convergence

$$T_{B0,E}^{a^*} = e^{\delta \hat{t}_{B0,E}^*} T_{B0,E}^{a}, \tag{9.34}$$

where $\delta t_{B0,E}^* = \left(\delta x^*, \delta y^*, \delta z^*, \delta \alpha_x^*, \delta \beta_y^*, \delta \gamma_z^*\right)^T$ represents the noise injected. Here $\delta x^*, \delta y^*$, and $\delta z^* \in \mathbf{U}[-0.1, 0.1]$ mm; $\delta \alpha_x^*, \delta \beta_y^*$ and $\delta \gamma_z^* \in \mathbf{U}[-0.0005, 0.0005]$ rad. Extensive simulations have been done to study the effect of the measurement noise and robot repeatability on the calibration results. In each simulation, two groups of 50 simulated end-effector poses are employed. One group is used to calibrate the robot, while the other group is used to verify the result of the calibration, i.e., to

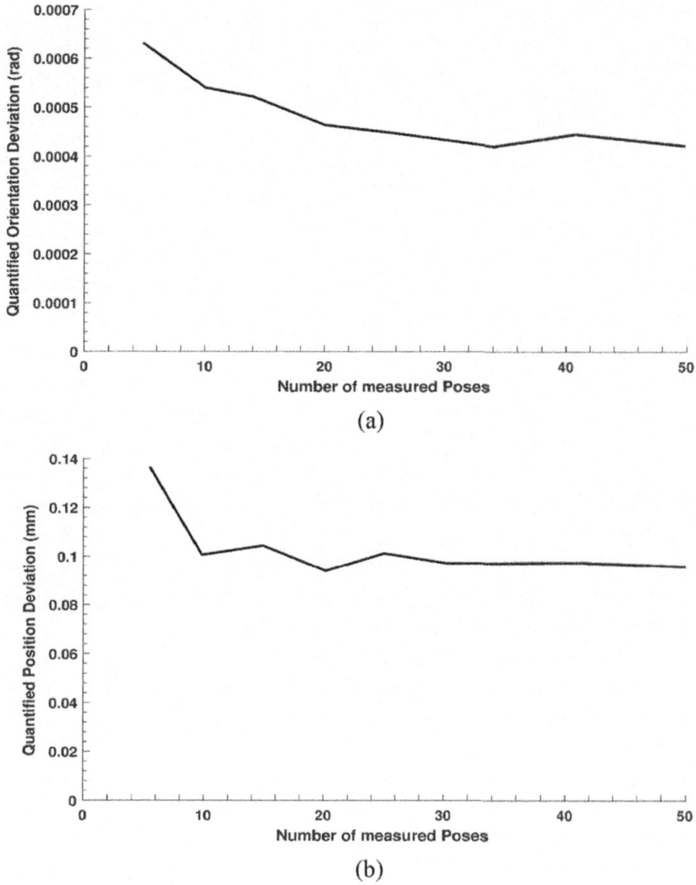

Fig. 9.6 Quantified deviation versus number of measurement postures

derive the quantified orientation and position deviations. From these simulations, it can be found that the calibrated initial poses of the module frames and the quantified orientation and position deviations become stable when the number of poses used for calibration is greater than 20. A typical simulation result is given in Fig. 9.6, in which the stable quantified orientation and position deviations are in the same order of magnitude as those of the injected noise.

References

1. R.O. Ambrose, *Design, Construction and Demonstration of Modular, Reconfigurable Robots* (University of Texas at Austin, U.S.A, 1991). PhD thesis
2. R.O. Ambrose, D. Tesar, Modular robot connection design. in *Proceedings of ASME Conference on Flexible Assembly Systems*, (Scottsdate, AZ, 1992), pp. 41–48
3. B. Benhabib, R. Cohen, M.G. Lipton, M.Q. Dai, Design of a rotary-joint-based modular robot. in *Proceedings of ASME Mechanism Conference*, (1990), pp. 239–243
4. I.M. Chen, G. Yang, Configuration independent kinematics for modular robots. in *Proceedings of IEEE Conference on Robotics and Automation*, (1996), pp. 1845–1849
5. R. Cohen, M.G. Lipton, M.Q. Dai, B. Benhabib, Conceptual design of a modular robot. ASME J. Mech. Des. **114**, 117–125 (1992)
6. T. Fukuda, S. Nakagawa, Dynamically reconfigurable robotic system. in *Proceedings of IEEE Conference on Robotics and Automation*, (Philadelphia, PA, April 1988), pp. 1581–1586
7. J. Han, W.K. Chung, S.H. Kim, Task based design of modular robot manipulator using efficient genetic algorithm. in *Proceedings of IEEE Conference on Robotics and Automation*, (New Mexico, USA, 1997), pp. 507–511
8. T. Matsumaru, Design and control of the modular robot system: Tomms. in *Proceedings of IEEE Conference on Robotics and Automation*, (Nagoya, Japan, May 1995), pp. 2125–2131
9. C.J.J. Paredis, H.B. Brown, R.W. Casciola, J.E. Moody, P.K. Khosla, A rapidly deployable manipulator system, in *International Workshop on Some Critical Issues in Robotics*. (Singapore, 1995), pp. 175–185
10. D. Schmitz, P. Khosla, T. Kanade, *The cmu reconfigurable modular manipulator system*. Technical Report CMU-RI-TR-88-7, (Carnegie Mellon University, 1988)
11. D. Tesar, M.S. Butler, A generalized modular architecture for robot structures. ASME J. Manuf. Rev. **2**(2), 91–117 (1989)
12. K.H. Wurst, The conception and construction of a modular robot system. in *International Symposium on Industrial Robotics*, (1986), pp. 37–44
13. B. Benhabib, G. Zak, M.G. Lipton, A generalized kinematic modeling method for modular robots. J. Rob. Syst. **6**(5), 545–571 (1989)
14. I.M. Chen, *Theory and Applications of Modular Reconfigurable Robotic Systems* (California Institute of Technology, CA, USA, 1994). PhD thesis

G. Yang and I.-M. Chen, *Modular Robots: Theory and Practice*,
Research on Intelligent Manufacturing,
https://doi.org/10.1007/978-981-16-5007-9

15. I.M. Chen, Yeo Song Huat, Chen Guang, G. Yang, Kernel for modular robot applications - automatic modeling techniques. Int. J. Rob. Res. **18**(2), 225–242 (1999)
16. I.M. Chen, G. Yang, Inverse kinematics for modular reconfigurable robots. in *IEEE Conference on Robotics and Automation*, (1998), pp. 1647–1652
17. L. Kelmar, P. Khosla, Automatic generation of kinematics for a reconfigurable modular manipulator system. in *Proceedings of IEEE Conference on Robotics and Automation*, (Philadelphia, PA, April 1988), pp. 663–668
18. G. Legnani, R. Riva, Kinematics of modular robots. in *Proceedings of World Congress on the Theory of Machines and Mechanisms*, (1987), pp. 1159–1162
19. I.M. Chen, G. Yang, Automatic generation of dynamics for modular robots with hybrid geometry. in *Proceedings of IEEE Conference on Robotics and Automation*, (1997), pp. 2288–2293
20. I.M. Chen, G. Yang, Automatic model generation for modular reconfigurable robot dynamics. ASME J. Dyn. Syst. Meas. Contr. **120**, 346–352 (1997)
21. I.M. Chen, G. Yang, Geometric formulation of modular reconfigurable robot kinematics and dynamics. in *Proceedings of the 2nd Asian Control Conference*, (1997), pp. 265–268
22. C.J.J. Paredis, *An Agent-Based Approach to the Design of Rapidly Deployable Fault Tolerant Manipulators* (Carnegie Mellon University, USA, 1996). PhD thesis
23. I.M. Chen, G. Yang, Kinematic calibration of modular reconfigurable robots using product-of-exponentials formula. J. Rob. Syst. **14**(11), 807–821 (1997)
24. C.T. Tan, G. Yang, I.M. Chen, A modular kinematic calibration algorithm for industrial robots: simulation and experiment. in *IASTED International Conference on Robotics and Manufacturing*, (Canada, 1998), pp. 231–234
25. G. Yang, I.M. Chen, Kinematic calibration of modular reconfigurable robots. in *Proceedings of International Conference on Control, Automation, Robotics, and Vision*, (Singapore, 1996), pp. 1845–1849
26. G. Yang, I.M. Chen, A novel kinematic calibration algorithm for modular reconfigurable robotic systems. in *Proceedings of IEEE Conference on Robotics and Automation*, (1997), pp. 3197–3202
27. I.M. Chen, On optimal configuration of modular reconfigurable robots. in *Proceedings of International Conference on Control, Automation, Robotics, and Vision*, (Singapore, 1996)
28. I.M. Chen, J.W. Burdick, Determining task optimal modular robot assembly configurations. in *Proceedings of IEEE Conference on Robotics and Automation*, (Nagoya, Japan, 1995), pp. 132–137
29. O. Chocron, P. Bidaud, Genetic design of 3d modular manipulator. in *Proceedings of IEEE Conference on Robotics and Automation*, (1997), pp. 223–228
30. T. Fukuda, S. Nakagawa, Y. Kawauchi, M. Buss, Structure decision method for self organizing robots based on cell structures-cebot. in *Proceedings of IEEE Conference on Robotics and Automation*, (Scottsdale, AZ, 1989), pp. 695–700
31. J.O. Kim, P. Khosla, A formulation for task-based design of robot manipulators. in *Proceedings of IEEE/RSJ International Conference on Intelligent Robots and Systems*, (1993), pp. 2310–2317
32. C.J.J. Paredis, P. Khosla, Kinematic design of serial link manipulators from task specifications. Int. J. Rob. Res. **12**(3), 274–287 (1993)
33. C.J.J. Paredis, P.K. Khosla, Synthesis Methodology for Task Based Reconfiguration of Modular Manipulator Systems, in *International Symposium on Robotics Research*, (Hidden Valley, PA, 1993)
34. G. Yang, I.M. Chen, Reduced DOF modular robot configurations. in *Proceedings of International Conference on Control, Automation, Robotics, and Vision*, (Singapore, 1998), pp. 1513–1517
35. C.J.J. Paredis and P.K. Khosla. Design of modular fault tolerant manipulators. in *Algorithmic Foundations of Robotics*, ed. by K. Goldberg, D. Halperin, J.-C. Latombe, R. Wilson, A.K. Peters (1995), pp. 371–383
36. C.J.J. Paredis, P.K. Khosla, Agent-based design of fault tolerant manipulators for satellite docking. in *Proceedings of IEEE Conference on Robotics and Automation*, (New Mexico, USA, 1997), pp. 3473–3480

37. N. Napp, S. Burden, E. Klavins. The statistical dynamics of programmed self-assembly. in *Robotics and Automation, 2006. ICRA 2006. Proceedings 2006 IEEE International Conference on*, (2006)
38. M.T. Tolley, J.D. Hiller, H. Lipson, *Evolutionary Design and Assembly Planning for Stochastic Modular Robots* (Springer, Berlin Heidelberg, 2011)
39. M. Yim, Y. Zhang, K. Roufas, D. Duff, C. Eldershaw, Connecting and disconnecting for chain self-reconfiguration with polybot. IEEE/ASME Trans. Mechatron. **7**(4), 442–451 (2003)
40. J. Zhao, X. Cui, Y. Zhu, S. Tang, A new self-reconfigurable modular robotic system ubot: Multi-mode locomotion and self-reconfiguration. in *2011 IEEE International Conference on Robotics and Automation (ICRA)*, (2011)
41. J. Zhao, X. Cui, Y. Zhu, S. Tang, Ubot: a new reconfigurable modular robotic system with multimode locomotion ability. *Industrial Robot*, volume 39(2):178–190(13), 2012
42. H. Kurokawa, K. Tomita, A. Kamimura, S. Kokaji, T. Hasuo, S. Murata, Distributed self-reconfiguration of m-tran iii modular robotic system. Int. J. Rob. Res. **27**(3–4), 373–386 (2008)
43. E.H. ØStergaard, K. Kassow, R. Beck, H.H. Lund, Design of the atron lattice-based self-reconfigurable robot. Auton. Robots **21**(2), 165–183 (2006)
44. W.M. Shen, M. Krivokon, H. Chiu, J. Everist, M. Rubenstein, J. Venkatesh, Multimode locomotion via superbot reconfigurable robots. Auton. Robots, **20**(2), 165–177 (2006)
45. J. Davey, N. Kwok, M. Yim, Emulating self-reconfigurable robots - design of the smores system. in *IEEE/RSJ International Conference on Intelligent Robots and Systems*, (2012)
46. T. Fukuda, T. Ueyama, F. Arai, Control strategy for a network of cellular robots. in *Proceedings of IEEE Conference on Robotics and Automation*, (Sacramento, CA, April 1991), pp. 1616–1621
47. R.J. Alattas, P. Sarosh, T.M. Sobh, Evolutionary modular robotics: survey and analysis. J. Intell. Rob. Syst. (2018)
48. B. Bollobás, *Graph Theory - An Introductory Course* (Springer-Verlag, 1979)
49. R.J. Wilson, J.J. Watkins, *Graphs - An Introductory Approach* (John Wiley & Sons Inc., New York, USA, 1989)
50. G. Chartrand, O.R. Oellermann, *Applied and Algorithmic Graph Theory* (McGraw-Hill Inc, USA, 1993)
51. S. Skiena, *Implementing Discrete Mathematics* (Addison-Wesley, Redwood City, CA, 1990)
52. L. Dobrjanskyj, F. Freudenstein, Some applications of graph theory to the structural analysis of mechanisms. ASME J. Eng. Ind. **89**, 153–158 (1967)
53. F. Harary, H. Yan, Logical foundations of kinematic chains: graphs, line graphs, and hypergraphs. ASME J. Mech. Des. **112**, 79–83 (1990)
54. L.S. Woo, Type synthesis of plane linkages. ASME J. Eng. Ind. **89**, 159–172 (1967)
55. H.-S. Yan, Y.-W. Hwang, The specialization of mechanisms. Mechanism and Machine Theory **26**(6), 541–551 (1991)
56. J. Denavit, R.S. Hartenberg, Kinematic notation for lower-pair mechanisms based on matrices. ASME Journal of Applied Mechanics **2**, 215–211 (1955)
57. J. Duffy, *Analysis of Mechanisms and Robot Manipualtors* (John Wiley & Sons, 1980)
58. C.H. Suh, C.W. Radcliffe, *Kinematics and Mechanisms Design* (John Wiley & Sons, 1978)
59. K.C. Gupta, Kinematic analysis of manipulator using the zero reference position description, in *The Kinematics of Robot Manipulators*. ed. by J.M. McCarthy. pp. (MIT Press, Cambridge, MA, 1987), pp. 3–11
60. R. Brockett, Robotic manipulators and the product of exponential formula, in *International Symposium in Math.* Theory of Network and Systems. (Beer Sheba, Israel, 1983), pp. 120–129
61. Z. Li, Geometrical considerations of robot kinematics. IEEE Trans. Robot. Autom. **5**(3), 139–145 (1990)
62. R. Murray, Z. Li, S.S. Sastry, *A Mathematical Introduction to Robotic Manipulation* (CRC Press, 1994)
63. F.C. Park, On the optimal kinematic design of spherical and spatial mechanisms. in *Proceedings of IEEE Conference on Robotics and Automation*, (Sacramento, CA, April 1991), pp. 1530–1535

64. F.C. Park, R.W. Brockett, Kinematic dexterity of robotic mechanisms. Int. J. Robot. Res. **13**(1), 1–15 (1994)
65. J.J. Craig, *Introduction to Robotics Mechanics and Control*, 2nd edn. (Addison-Wesley, New York, USA, 1989)
66. P.K. Khosla, C.P. Neuman, F.B. Prinz, An algorithm for seam tracking applications. Int. J. Robot. Res. **4**(1), 27–41 (1985)
67. C.S.G. Lee, Robot arm kinematics, dynamics, and control. IEEE Comput. **15**(12), 62–68 (1982)
68. R.P. Paul, *Robot Manipulators: Mathematics, Programming, and Control* (MIT Press, Cambridge, MA, 1981)
69. R.P. Paul, B. Shimano, G.E. Mayer, Kinematic control equation for simple manipulators. IEEE Trans. Syst. Man Cybern. **11**(6), 80–86 (1981)
70. J. Angeles, Numerical solution to the input output displacement equation of general $7R$ spatial mechanism. in *Proceedings of the Fifth World Congress and Mechanisms*, (1979), pp. 1008–1011
71. J. Duffy, C. Crane, A displacement analysis of the general spacial $7r$ mechanism. Mech. Mach. Theory **15**, 153–169 (1980)
72. H.Y. Lee, C.G. Liang, Displacement analysis of the general serial 7-link 7-r mechanism. Mech. Mach. Theory **23**, 219–226 (1988)
73. H.Y. Lee, C.G. Liang, A new vector theory for the analysis of spatial mechanisms. Mech. Mach. Theory **23**, 209–217 (1988)
74. H.Y. Lee, C.F. Reinholtz, Inverse kinematics of serial-chain manipulators. ASME J. Mech. Des. **118**, 396–404 (1996)
75. J. Denavit, R.S. Hartenberg, An iterative method for the displacement analysis of special mechanisms. ASME J. Appl. Mech. 309–314 (1964)
76. K.C. Gupta, K. Kazerounian, Improved numerical solution of inverse kinematics of robots. in *Proceedings of IEEE Conference on Robotics and Automation*, (Philadelphia, PA, April 1988), pp. 743–748
77. J. Lenar*checkcicheckc*. , An efficient numerical approach for calculating the inverse kinematics for robot manipulators. Robotica 21–26 (1985)
78. R. Manseur, K.L. Doty, A fast algorithm for inverse kinematic analysis of robot manipulators. Int. J. Robot. Res. **4**(2), 21–37 (1988)
79. D.R. Meldrum, G. Rodriguez, G.F. Franklin, An order(n) recursive inversion of the jacobian for an n-link serial manipulator. in *Proceedings of IEEE Conference on Robotics and Automation*, (Sacramento, CA, 1991), pp. 1175–1180
80. T.C. Yih, Y. Youm, Matrix solution for the inverse kinematics of robots. in *ASME International Conference*, (1990), pp. 371–375
81. J. Angeles, On the numerical solution of the inverse kinematics problem. Int. J. Robot. Res. **4**(2), 21–37 (1985)
82. A.A. Goldenberg, D.L. Lawrence, A generalized solution to the inverse kinematics of robotic manipulators. ASME J. Dyn. Syst. Meas. Contr. **107**, 103–106 (1985)
83. D. Manocha, J.F. Canny, Real time inverse of the general $6r$ manipulators. in *Proceedings of IEEE Conference on Robotics and Automation*, (Nice France, May 1992), pp. 383–389
84. M. Raghavan, B. Roth, Kinematic analysis of the $6R$ manipulator of general gepmetry. in *International Symposium on Robotics Research*, (1989), pp. 314–320
85. C. Wampler, A.P. Morgen, Solving the $6r$ inverse position problem using a geometric-case solution methodology. Mech. Mach. Theory **26**, 91–106 (1991)
86. R. Featherstone, Explicit modeling of general task spaces for inverse kinematics, *Advances in Robot Kinematics and Computational Geometry* (MIT Press, Cambridge, 1994), pp. 301–308
87. F.C. Park, Computational aspect of manipulators via product of exponential formula for robot kinematics. IEEE Trans. Autom. Control **39**(9), 643–647 (1994)
88. F.C. Park, J.E. Bobrow, S.R. Ploen, A lie group formulation of robot dynamics. Int. J. Robot. Res. **14**(6), 609–618 (1995)
89. R. Bernhardt, S.L. Albright, *Robot Calibration* (Chapman and Hall, London; New York, 1993)

90. J.M. Hollerbach, A survey of kinematic calibration, *The Robotics Review I* (MIT Press, Cambridge, 1989), pp. 207–242
91. B.W. Mooring, Z.S. Roth, M.R. Driels, *Fundamentals of Manipulator Calibration* (John Wiley and Sons, 1991)
92. Z. Roth, B.W. Mooring, B. Ravani, An overview of robot calibration. IEEE J. Robot. Autom. **3**(5), 377–386 (1987)
93. H.W. Stone, *Kinematic Modeling, Identification, and Control of robotic Manipulators*, (Kluwer Academic Publishers, 1987)
94. S.A. Hayati, Robot arm geometric parameter estimation. in *Proceedings of IEEE Conference on Decision and Control*, (1983), pp. 1477–1483
95. S.A. Hayati, M. Mirmirani, Improving the absolute positioning accuracy of robot manipulators. J. Robot. Syst. **112**(2), 397–413 (1985)
96. S.A. Hayati, G.P. Roston, Inverse kinematic solution for near-simple robots and its application to robot calibration, *Recent Trends in Robotics: Modeling, Control, and Education* (Elsevier Science, 1986), pp. 41–50
97. S.A. Hayati, G.P. Roston, Robot geometry calibration. in *Proceedings of IEEE Conference on Robotics and Automation*, (Philadelphia, PA, April 1988), pp. 947–951
98. D.E. Whitney, C.A. Lozinski, J.M. Rourke, Industrial robot forward calibration method and results. J/ Dyn. Syst. **108**, 1–8 (1986)
99. C. Wu, The kinematic error model for the design of robot manipulators. in *Proceedings of the American Control Conference*, (1983), pp. 497–502
100. K. Kazerounian, G.Z. Qian, Kinematic calibration of robotics manipulators. in *ASME Design Technology Conferences: Biennial Mechanisms Conference*, (1988), pp. 261–266
101. K. Kazerounian, G.Z. Qian, Kinematic calibration of robotic manipulator. ASME J. Mech. Trans. Autom. Des. **111**, 482–487 (1989)
102. B.W. Mooring, The effect of joint axis misalignment on robot positioning accuracy. in *Proceedings of Computers in Engineering Conference and Exhibit*, (1983), pp. 151–155
103. B.W. Mooring, An improved method for identifying the kinematic parameters in a six axis robot. in *Proceedings of Computers in Engineering Conference and Exhibit*, (1984), pp. 79–84
104. H. Zhuang, Z.S. Roth, Robot calibration using the cpc error model. J. Robot. Computer-Integrated Manuf **9**(3), 227–237 (1992)
105. H. Zhuang, Z.S. Roth, A linear solution to the kinematic parameter identification of robot manipulator. IEEE Trans. Robot. Autom. **9**(2), 174–185 (1993)
106. H. Zhuang, Z.S. Roth, H. Fumio, A complete and parametrically continuous kinematic model for robot manipulators. IEEE Trans. Robot. Autom. **8**(4), 451–463 (1992)
107. K. Okamura, F.C. Park, Kinematic calibration and the product of exponential formula. Robotica **14**, 415–421 (1996)
108. F.C. Park, K. Okamura, Kinematic calibration and the product of exponential formula, *Advances in Robot Kinematics and Computational Geometry* (MIT Press, Cambridge, 1994), pp. 119–128
109. C.T. Tan, G. Yang, I.M. Chen, A generic kinematic calibration model for the open chain robots. in *Proceedings of International Conference on Control, Automation, Robotics, and Vision*, (Singapore, 1998), pp. 1503-1507
110. J.M. Hollerbach, D.J. Bernett, Automatic kinematic calibration using a motion tracking system. in *Modeling and Control of Robotic Manipulators and Manufacturing Processes*, (1987), pp. 93–100
111. R.P. Judd, A.B. Knasinski, A technique to calibrate industrial robots with experimental verification. in *Proceedings of IEEE Conference on Robotics and Automation*, (Raleigh, NC, March 1987), pp. 351–357
112. K. Sugimoto, T. Okada, Compensation of positioning errors caused by geometric deviations in robot system, in *International Symposium on Robotics Research*, ed. by H. Hanafusa, H. Inoue (MIT Press, Cambridge, MA, 1985), pp. 231–236
113. W.K. Veitschegger, C.H. Wu, Robot accuracy analysis based on kinematics. IEEE Trans. Robot. Autom. **2**, 171–179 (1986)

114. F.C. Park, J.E. Bobrow, A recursive algorithm for robot dynamics using lie groups. in *Proceedings of IEEE Conference on Robotics and Automation*, (San Diego, CA, 1994), pp. 1535–1540
115. W.W. Armstrong, Recursive solution to the equation of motion in an n-link manipulator. in *Proceedings of the Fifth World Congress on Mechanisms and Machine Theory*, (1979), pp. 1134–1346
116. J.Y.S. Luh, M.W. Walker, R.P. Pual, On-line computational scheme for mechanical manipulators. ASME J. Dyn. Syst. Meas. Contr. **102**, 103–110 (1980)
117. R.L. Huston, C.E. Passerrello, M.W. Harlow, Dynamics of multirigid-body system. ASME J. Appl. Mech. **45**(4), 889–894 (1978)
118. T.R. Kane, D.A. Levinson, The use of kane's dynamic equation in robotics. Int. J. Robot. Res. **2**(3), 1–21 (1983)
119. C.S.G. Lee, B.H. Lee, R. Nigam, Development of the generalized d'alembert equation of motion for mechanical design. in *Proceedings of IEEE Conference on Decision and Control*, (1983), pp. 9998–9999
120. J.M. Hollerbach, A recursive lagrangian formulation of manipulator dynamics and a comparative study of dynamics formulation complexity. IEEE Trans Syst Man Cybern **10**, 730–736 (1980)
121. R. Brockett, A. Stokes, F.C. Park, A geometric formulation of the dynamic equation describing kinematic chains. in *Proceedings of IEEE Conference on Robotics and Automation*, (Altanta, GA, May 1993), pp. 637–641
122. F.C. Park, A coordinate-free description of robot dynamics. in *Proceedings of IEEE Conference on Robotics and Automation*, (New Mexico, USA, 1997), pp. 2282–2287
123. M.W. Spong, Remarks on robot dynamics: canonical transformation and Riemannian geometry. in *Proceedings of IEEE Conference on Robotics and Automation*, (Nice France, May 1992), pp. 554–559
124. J. Russakow, O. Khatib, S.M. Rock, Extended operational space formulation for serial-to-parallel chain (branching) manipulators. in *Proceedings of IEEE Conference on Robotics and Automation*, (Nagoya, Japan, 1995), pp. 1056–1061
125. R. Featherstone, *Robot Dynamics Algorithms* (Kluwer Academic Publishers, 1987)
126. S.R. Ploen, F.C. Park, Coordinate-invariant algorithms for robot dynamics, in *International Workshop on Some Critical Issues in Robotics*, (Singapore, 1995), pp. 115–128
127. N. Deo, *Graph Theory with Applications to Engineering and Computer Science*, (Prentice-Hall, 1974)
128. I.M. Chen, G. Yang, An evolutionary algorithm for the reduction of DOFs in modular reconfigurable robots. in *Proceedings of the ASME Dynamic System and Control Division*, DSC-64, (1998) pp. 759–766
129. G. Yang, I.M. Chen, Task-based optimization of modular robot configurations - MDOF approach. Mech. Mach. Theory (1999)
130. B. Roth, Performance evaluation of manipulators from a kinematic point of view. in *NBS Special Publication 495*, (National Bureau of Standards, 1976), pp. 39–61
131. B. Paden, S. Sastry, Optimal kinematic design of 6r manipulators. Int. J. Robot. Res. **7**(2), 43–61 (1988)
132. F.C. Park, Optimal robot design and differential geometry. ASME Special 50th Anniversary Design Issue **117**, 87–91 (1995)
133. K.C. Gupta, B. Roth, Design considerations for manipulator workspace. ASME J. Mech. Des. **104**, 704–711 (1982)
134. K.C. Gupta, On the nature of robot workspace. Int. J. Robot. Res. **5**(2), 112–121 (1986)
135. F. Freudenstein, E. Primrose, On the analysis and sythesis of the workspace of a three-link, turning pair connected robot arm. ASME J. Mech. Trans. Autom. Des. **106**, 365–370 (1984)
136. W.C. Lin, J.B. Ross, M. Ziegler, Semiautomatic calibration of robot manipulator for visual inspection task. J. Robot. Syst. **3**, 19–39 (1986)
137. D.C.H. Yang, T.W. Lee, On the workspace of mechanical manipulator. ASME J. Mech. Transm. Autom. Des. **105**, 62–69 (1983)

138. L.-W. Tsai, A.P. Morgan, Solving the kinematics of the most general 6 and 5 dof manipulators by continuation methods. ASME J. Mech. Transm. Autom. Des. **107**, 189–200 (1985)
139. A. Kumar, K.J. Waldron, The workspaces of a mechanical manipulator. ASME J. Mech. Des. **103**, 665–672 (1981)
140. F.C. Yang, E.J. Hang, Numerical analysis of the kinematic working capability of mechanisms. ASME J. Mech. Des. **116**, 111–118 (1994)
141. J.K. Salisbury, J.J. Craig, Articulated hands: force control and kinematic issues. Int. J. Robot. Res. **1**(1), 4–17 (1982)
142. T. Yoshikawa, Manipulability of robotic mechanisms. Int. J. Robot. Res. **4**(2), 3–9 (1985)
143. J. Angeles, The design of isotropic manipulator architectures in the presence of redundancies. Int. J. Robot. Res. **11**(3), 196–201 (1992)
144. C. Gosselin, J. Angeles, A global performance index for the kinematic optimization of robotic manipulators. ASME J. Mech. Des. **113**, 220–226 (1991)
145. C.A. Klein, B.E. Blaho, Dexterity measures for the design and control of kinematically redundant manipulators. Int. J. Robot. Res. **6**(2), 72–83 (1987)
146. J.M. Hollerbach, *Optimum Kinematics Design of a Seven Degree of Freedom Manipulator*, (MIT Press, Cambridge, MA, 1985)
147. S.L. Chiu, Kinematic characterization of manipulators: an approach to defining optimality. in *Proceedings of IEEE Conference on Robotics and Automation*, (Philadelphia, PA, 1988), pp. 828–833
148. K.V.D. Doel, D.K. Pai, Performance measures for robot manipulators: a unified approach. Int. J. Robot. Res. **15**(1), 92–111 (1996)
149. J.O. Kim, P. Khosla, Design of space shuttle tile servicing robot: an application of task based kinematic design. in *Proceedings of IEEE Conference on Robotics and Automation*, (Altanta, GA, 1993), pp. 867–874
150. Z. Michalewicz, *Genetic Algorithms + Data Structures = Evolution Programs*, 2nd edn. (Springer-Verlag, 1994)
151. D. Stewart, A platform with six degrees of freedom. Proc. Inst. Mech. Eng. **180**(1), 371–386 (1965)
152. D. Chablat, P. Wenger, Architecture optimization of a 3-dof translational parallel mechanism for machining applications, the orthoglide. IEEE Trans. Robot. Autom. **19**(3), 403–410 (2003)
153. D. Lazard, J.P. Merlet, The (true) stewart platform has 12 configurations. in *Proceedings of the 1994 IEEE International Conference on Robotics and Automation*, (IEEE, 1994), pp. 2160–2165
154. I. Zabalza, J. Ros, J.J. Gil, J.M. Pintor, J.M. Jimenez, Tri-scott. a new kinematic structure for a 6-dof decoupled parallel manipulator. in *Proceedings of the workshop on fundamental issues and future research directions for parallel mechanisms and manipulators*, pp. 12–15
155. J. Kim, F.C. Park, S.J. Ryu, J. Kim, J.C. Hwang, C. Park, C.C. Iurascu, Design and analysis of a redundantly actuated parallel mechanism for rapid machining. IEEE Trans. Robot. Autom. **17**(4), 423–434 (2001)
156. G. Coppola, D. Zhang, K. Liu, A 6-dof reconfigurable hybrid parallel manipulator. Robot. Computer-Integrated Manuf. **30**(2), 99–106 (2014)
157. Lu. Yi, P. Wang, Z. Hou, H. Bo, C. Sui, J. Han, Kinetostatic analysis of a novel 6-dof 3ups parallel manipulator with multi-fingers. Mech. Mach. Theory **78**, 36–50 (2014)
158. D. Glozman, M. Shoham, Novel 6-dof parallel manipulator with large workspace. Robotica **27**(6), 891 (2009)
159. C. Innocenti, V. Parenti-Castelli, Direct position analysis of the stewart platform mechanism. Mech. Mach. Theory **25**(6), 611–621 (1990)
160. L. Notash, R.P. Podhorodeski, Complete forward displacement solutions for a class of three branch parallel manipulators. J. Robot. Syst. **11**(6), 471–485 (1994)
161. K. Cleary, T. Brooks, Kinematic analysis of a novel 6-dof parallel manipulator. in *[1993] Proceedings IEEE International Conference on Robotics and Automation*, (IEEE, 1993), pp. 708–713

162. C. Gosselin, J. Angeles, Singularity analysis of closed-loop kinematic chains. IEEE Trans. Robot. Autom. **6**(3), 281–290 (1990)
163. K.H. Hunt, *Kinematic Geometry of Mechanisms*, vol. 7 (Oxford University Press, USA, 1978)
164. V. Kumar, Instantaneous kinematics of parallel-chain robotic mechanisms. (1992)
165. J.P. Merlet, Singular configurations of parallel manipulators and grassmann geometry. Int. J. Robot. Res. **8**(5), 45–56 (1989)
166. H. Zhou, Y. Cao, B. Li, J. Yu, M. Liu, Singularity kinematics principle and singularity analysis of 6–3 stewart manipulators for special orientations. in *2010 Chinese Control and Decision Conference*, (IEEE, 2010), pp. 4336–4341
167. K. Malakuti, R.E. Caflisch, M. Siegel, A. Virodov, Detection of complex singularities for a function of several variables. IMA J. Appl. Math. **78**(4), 714–728 (2013)
168. M.S. Islam, H. Koguchi, Analysis of intensity of singularity at a vertex in 3d transversely isotropic piezoelectric bimaterial joints by boundary element method. in *2013 International Conference on Informatics, Electronics and Vision (ICIEV)*, (IEEE, 2013), pp. 1–6
169. I.A. Bonev, J. Ryu, A new approach to orientation workspace analysis of 6-dof parallel manipulators. Mech. Mach. Theory **36**(1), 15–28 (2001)
170. Eugene F. Fichter, A stewart platform-based manipulator: general theory and practical construction. Int. J. Robot. Res. **5**(2), 157–182 (1986)
171. O. Masory, J. Wang, Workspace evaluation of stewart platforms. Adv. Robot. **9**(4), 443–461 (1994)
172. T. Arai, Development of a new parallel manipulator with fixed linear actuator, in*Proc. of Japan/USA Symposium on Flexible Automation*, (1995), pp. 145–149
173. E. Ottaviano, M. Ceccarelli, Optimal design of capaman (cassino parallel manipulator) with a specified orientation workspace. Robotica **20**(2), 159 (2002)
174. J.P. Merlet, Détermination de l'espace de travail d'un robot parallèle pour une orientation constante. Mech. Mach. Theory **29**(8), 1099–1113 (1994)
175. R.P. Podhorodeski, K.H. Pittens, A class of parallel manipulators based on kinematically simple branches. (1994)
176. Z. Liu, R. Zhang, W. Wang, X. Ma, R. You, Z. Li, Robot workspace reconstructing solving method based on image engine. in *2017 International Conference on Industrial Informatics-Computing Technology, Intelligent Technology, Industrial Information Integration (ICIICII)*, (IEEE, 2017) pp. 181–185
177. Z. Zhao, S. He, Y. Zhao, C. Xu, Q. Wu, Z. Xu, Workspace analysis for a 9-dof hyper-redundant manipulator based on an improved monte carlo method and voxel algorithm. in *2018 IEEE International Conference on Mechatronics and Automation (ICMA)*, (IEEE, 2018), pp. 637–642
178. A. Peidró, Ó. Reinoso, A. Gil, J.M. Marín, L. Payá, A method based on the vanishing of self-motion manifolds to determine the collision-free workspace of redundant robots. Mech. Mach. Theory **128**, 84–109 (2018)
179. R.P. Podhorodeski, K.H. Pittens, A class of parallel manipulators based on kinematically simple branches. Trans. ASME J. Mech. Des **116**(3), 908–914 (1994)
180. G. Yang, I.M. Chen, W.K. Lim, S.H. Yeo, Kinematic design of modular reconfigurable in-parallel robots. Auton. Robots **10**(1), 83–89 (2001)
181. G. Yang, I.M. Chen, W.K. Lim, H.Y. Song, Design and kinematic analysis of modular reconfigurable parallel robots. in *IEEE International Conference on Robotics and Automation*
182. G. Yang, Kinematics, dynamics, calibration, and configuration optimization of modular reconfigurable robots. (1999)
183. D. Lazard, J.P. Merlet, The (true) stewart platform has 12 configurations. in *Proceedings of the 1994 IEEE International Conference on Robotics and Automation*, (IEEE, 1994), pp. 2160–2165
184. I. Zabalza, J. Ros, J.J. Gil, J.M. Pintor, J.M. Jimenez Tri-scott. a new kinematic structure for a 6-dof decoupled parallel manipulator. in *Proceedings of the workshop on fundamental issues and future research directions for parallel mechanisms and manipulators*, (2002), pp. 12–15

185. I.G. Yang, M. Chen, W. Lin, J. Angeles, Singularity analysis of three-legged parallel robots based on passive-joint velocities. IEEE Trans. Robot. Autom. **17**(4), 413–422 (2001)
186. R.W. Brockett, Robotic manipulators and the product of exponentials formula. in *Mathematical theory of networks and systems*, (Springer, 1984), pp. 120–129
187. R.M. Murray, S.Z. Li, S. Sastry, S.S. Sastry, A mathematical introduction to robotic manipulation. CRC Press (1994)
188. C.F. Park, Computational aspects of the product-of-exponentials formula for robot kinematics. IEEE Trans. Autom. Contr. **39**(3), 643–647 (1994)
189. C.M. Gosselin, J. Sefrioui, Polynomial solutions for the direct kinematic problem of planar three-degree-of-freedom parallel manipulators. in *International Conference on Advanced Robotics*
190. O. Ma, J. Angeles, Architecture singularities of parallel manipulators. Int. J. Robot. Autom. **7**(1), 23–29 (1992)
191. G. Yang, W. Chen, I.M. Chen, A geometrical method for the singularity analysis of 3-rrr planar parallel robots with different actuation schemes. in *IEEE/RSJ International Conference on Intelligent Robots and Systems*
192. C. Gosselin, J. Angeles, Singularity analysis of closed-loop kinematic chains. IEEE Trans. Robot. Autom. **6**(3), 281–290 (1990)
193. L.W. Tsai, *Robot Analysis: The Mechanics of Serial and Parallel Manipulators* (John Wiley and Sons, 1999)
194. G. Yang, I.M. Chen, Equivolumetric partition of solid spheres with applications to orientation workspace analysis of robot manipulators. IEEE Trans. Robot. **22**, 869–879 (2006)
195. C.M. Gosselin, Determination of the workspace of 6-dof parallel manipulators. J. Mech. Des. **112**(3), 331–336 (1990)
196. J. Angeles, C.S. López-Cajún, Kinematic isotropy and the conditioning index of serial robotic manipulators. Int. J. Robot. Res. **11**(6), 560–571 (1992)
197. Rigelsford J, Fundamentals of robotic mechanical systems. Industrial Robot: An International Journal (1973)
198. C.M. Gosselin, L. Perreault, C. Vaillancourt, Simulation and computer-aided design of spherical parallel manipulators. in *Oceans*
199. S.J. Ryu, J. Kim, F.C. Park, J. Kim, Design and performance analysis of a parallel mechanism-based universal machining center. in *Proceedings of 3rd International Conference on Advanced Mechatronics*, (1998), pp. 359–363
200. F. Badano, M. Bétemps, R. Clavel, A. Jutard, M. Hongler, Evaluation of exploration strategies in robotic assembly. IFAC Proceedings Volumes **27**(14), 63–68 (1994)
201. M. Russo, X. Dong, A calibration procedure for reconfigurable gough-stewart manipulators. Mech Mach Theory 103920 (2020)
202. J.-S. Gao, L. Cheng, Y.S. Zhao, Kinetic self-calibration of novel 5-dof parallel machine tool. Comput. Integr. Manufactured Manuf. Syst. **13**(4), 738 (2007)
203. T. Sun, B. Lian, S. Yang, Y. Song, Kinematic calibration of serial and parallel robots based on finite and instantaneous screw theory. IEEE Trans Robot **36**(3), 816–834 (2020)
204. I. Dressler, A. Robertsson, R. Johansson, Automatic kinematic calibration of a modular gantry-tau parallel robot from a kinematics point of view. in *2008 IEEE International Conference on Robotics and Automation*, (IEEE, 2008), pp. 1282–1287
205. G. Yang, I.M. Chen, W.K. Lee, S.H. Yeo, Self-calibration of three-legged modular reconfigurable parallel robots based on leg-end distance errors. Robotica **19**(2), 187 (2001)
206. Y.J. Chiu, M.H. Perng, Self-calibration of a general hexapod manipulator using cylinder constraints. Int. J. Mach. Tools Manuf. **43**(10), 1051–1066 (2003)
207. Y. Meng, H. Zhuang, Autonomous robot calibration using vision technology. Robot. Computer-Integrated Manuf. **23**(4), 436–446 (2007)
208. D. Guanglong, P. Zhang, Online robot calibration based on vision measurement. Robot. Computer-Integrated Manuf. **29**(6), 484–492 (2013)
209. Y.J. Chiu, M.H. Perng, Self-calibration of a general hexapod manipulator with enhanced precision in 5-dof motions. Mech Mach Theory **39**(1), 1–23 (2004)

210. Y. Chen, F. Xie, X. Liu, Y. Zhou, Error modeling and sensitivity analysis of a parallel robot with scara (selective compliance assembly robot arm) motions. Chin. J. Mech. Eng. **27**(4), 693–702 (2014)
211. H. Wang, T. Gao, J. Kinugawa, K. Kosuge, Finding measurement configurations for accurate robot calibration: validation with a cable-driven robot. IEEE Trans Robot **33**(5), 1156–1169 (2017)
212. A. Müller, M. Ruggiu, Self-calibration of redundantly actuated pkm exploiting kinematic landmarks, *Computational Kinematics*, (Springer, 2014), pp. 93–102
213. G. Ecorchard, P. Maurine, Self-calibration of delta parallel robots with elastic deformation compensation. in *2005 IEEE/RSJ International Conference on Intelligent Robots and Systems*, (IEEE, 2005), pp. 1283–1288
214. M. de Campos Porath, L.A.F. Bortoni, R. Simoni, J.S. Eger. Offline and online strategies to improve pose accuracy of a stewart platform using indoor-gps. Precis. Eng. **63**, 83–93 (2020)
215. G. Chen, L. Kong, Q. Li, H. Wang, A simple two-step geometric approach for the kinematic calibration of the 3-prs parallel manipulator. Robotica **37**(5), 837–850 (2019)
216. H. Zhuang, Self-calibration of parallel mechanisms with a case study on stewart platforms. IEEE Trans. Robot. Autom. **13**(3), 387–397 (1997)
217. H. Zhuang, Self-calibration of parallel mechanisms with a case study on stewart platforms. IEEE Trans. Robot. Autom. **13**(3), 387–397 (1997)
218. H. Zhuang, L. Liu, Self-calibration of a class of parallel manipulators. in *Proceedings of IEEE International Conference on Robotics and Automation*, vol. 2 (IEEE, 1996), pp. 994–999
219. A. Nahvi, J.M. Hollerbach, V. Hayward, Calibration of a parallel robot using multiple kinematic closed loops. in *Proceedings of the 1994 IEEE International Conference on Robotics and Automation*, (IEEE, 1994), pp. 407–412
220. D.J. Bennett, J.M. Hollerbach et al. Autonomous calibration of single-loop closed kinematic chains formed by manipulators with passive endpoint constraints. IEEE Trans. Robot. Autom. **7**(5), 597–606 (1991)
221. C.W. Wampler, J.M. Hollerbach, T. Arai, An implicit loop method for kinematic calibration and its application to closed-chain mechanisms. IEEE Trans. Robot. Autom. **11**(5), 710–724 (1995)
222. C.G. Iurascu, F.C. Park, Geometric algorithms for closed chain kinematic calibration. in *Proceedings 1999 IEEE International Conference on Robotics and Automation (Cat. No. 99CH36288C)*, vol. 3, (IEEE, 1999), pp. 1752–1757
223. L. Notash, R.P. Podhorodeski, Kinematic calibration of parallel manipulators. in *IEEE International Conference on Systems, Man and Cybernetics. Intelligent Systems for the 21st Century*, vol. 4 (IEEE, 1995), pp. 3310–3315
224. F.C. Park, K. Okamura, Kinematic calibration and the product of exponentials formula, *Advances in robot kinematics and computational geometry* (Springer, 1994), pp. 119–128
225. I.M. Chen, G. Yang, Kinematic calibration of modular reconfigurable robots using product-of-exponentials formula. J. Rob. Syst. **14**(11), 807–821 (1997)
226. G. Yang, I.M. Chen, A novel kinematic calibration algorithm for reconfigurable robotic systems. in *Proceedings of International Conference on Robotics and Automation*, vol. 4 (IEEE, 1997), pp. 3197–3202
227. H. Zhuang, Z.S. Roth, R. Sudhakar, Simultaneous robot/world and tool/flange calibration by solving homogeneous transformation equations of the form ax= yb. IEEE Trans. Robot. Autom. **10**(4), 549–554 (1994)
228. F.C. Park, B.J. Martin, Robot sensor calibration: solving ax= xb on the euclidean group. IEEE Trans. Robot. Autom. **10**(5), 717–721 (1994)
229. F. Dornaika, R. Horaud, Simultaneous robot-world and hand-eye calibration. IEEE Trans. Robot. Autom. **14**(4), 617–622 (1998)
230. M. Shah, Solving the robot-world/hand-eye calibration problem using the kronecker product. J. Mech. Robot. **5**(3) (2013)
231. Aiguo Li, Lin Wang, Wu Defeng, Simultaneous robot-world and hand-eye calibration using dual-quaternions and kronecker product. International Journal of Physical Sciences **5**(10), 1530–1536 (2010)

232. Jan Heller, Didier Henrion, and Tomas Pajdla. Hand-eye and robot-world calibration by global polynomial optimization. In *2014 IEEE international conference on robotics and automation (ICRA)*, pages 3157–3164. IEEE, 2014
233. Wu Liao, Hongliang Ren, Finding the kinematic base frame of a robot by hand-eye calibration using 3d position data. IEEE Transactions on Automation Science and Engineering **14**(1), 314–324 (2016)
234. Guilin Yang, I-Ming Chen, Song Huat Yeo, and Wee Kiat Lim. Simultaneous base and tool calibration for self-calibrated parallel robots. *Robotica*, 20(4):367, 2002
235. Floris Ernst, Lars Richter, Lars Matthäus, Volker Martens, Ralf Bruder, Alexander Schlaefer, Achim Schweikard, Non-orthogonal tool/flange and robot/world calibration. The International Journal of Medical Robotics and Computer Assisted Surgery **8**(4), 407–420 (2012)
236. Junhyoung Ha, Donghoon Kang, Frank C. Park, A stochastic global optimization algorithm for the two-frame sensor calibration problem. IEEE Transactions on Industrial Electronics **63**(4), 2434–2446 (2015)
237. Lars Richter, Floris Ernst, Alexander Schlaefer, Achim Schweikard, Robust real-time robot-world calibration for robotized transcranial magnetic stimulation. The International Journal of Medical Robotics and Computer Assisted Surgery **7**(4), 414–422 (2011)
238. Wei Li, Mingli Dong, Lu Naiguang, Xiaoping Lou, Peng Sun, Simultaneous robot-world and hand-eye calibration without a calibration object. Sensors **18**(11), 3949 (2018)
239. Zijian Zhao, Simultaneous robot-world and hand-eye calibration by the alternative linear programming. Pattern Recognition Letters **127**, 174–180 (2019)
240. Yang Li, Junlei Hu, Baoxin Tao, Dedong Yu, Yihan Shen, Shengchi Fan, Yiqun Wu, and Xiaojun Chen. Automatic robot-world calibration in an optical-navigated surgical robot system and its application for oral implant placement. *International Journal of Computer Assisted Radiology and Surgery*, pages 1–8, 2020
241. Guilin Yang, I-Ming Chen, Wee Kiat Lim, and Song Huat Yeo. Design and kinematic analysis of modular reconfigurable parallel robots. In *Proceedings 1999 IEEE International Conference on Robotics and Automation (Cat. No. 99CH36288C)*, volume 4, pages 2501–2506. IEEE, 1999
242. M.L. Curtis, *Matrix Groups*, 2nd edn. (Springer-Verlag, New York, USA, 1984)
243. W. Stephen, *Mathematica: A System for Doing Mathematics by Computers*, 2nd edn. (Addison-Wesley, USA, 1991)

The manufacturer's authorised representative in the EU is Springer Nature Customer Service Centre GmbH, Europaplatz 3, 69115 Heidelberg, Germany. If you have any concerns regarding our products, please contact ProductSafety@springernature.com

Printed and bound by CPI Group (UK) Ltd, Croydon, CR0 4YY
30/09/2026
02225330-0004